FULLERENE NANO-WHISKERS

Edited by Kun'ichi Miyazawa

FULLERENE NANO-WHISKERS

PAN STANFORD PUBLISHING

Published by

Pan Stanford Publishing Pte. Ltd.
Penthouse Level, Suntec Tower 3
8 Temasek Boulevard
Singapore 038988

Email: editorial@panstanford.com
Web: www.panstanford.com

British Library Cataloguing-in-Publication Data
A catalogue record for this book is available from the British Library.

Fullerene Nanowhiskers

Copyright © 2012 Pan Stanford Publishing Pte. Ltd.

All rights reserved. This book, or parts thereof, may not be reproduced in any form or by any means, electronic or mechanical, including photocopying, recording or any information storage and retrieval system now known or to be invented, without written permission from the publisher.

For photocopying of material in this volume, please pay a copying fee through the Copyright Clearance Center, Inc., 222 Rosewood Drive, Danvers, MA 01923, USA. In this case permission to photocopy is not required from the publisher.

ISBN 978-981-4241-85-4 (Hardcover)
ISBN 978-981-4241-63-2 (eBook)

Printed in the USA

Contents

1. Introduction to Fullerene Nanowhiskers 1
   Kun'ichi Miyazawa

2. Growth, Structures, and Mechanical Properties of C_{60} Nanowhiskers 25
   Masaru Tachibana

3. Investigation of the Growth Mechanism of C_{60} Fullerene Nanowhiskers 43
   Kayoko Hotta and Kun'ichi Miyazawa

4. Preparation and Characterization of Fullerene Derivatives and Their Nanowhiskers 53
   Shigeo Nakamura, Kun'ichi Miyazawa, and Tadahiko Mashino

5. Vertically Aligned C_{60} Microtube Array 63
   Cha Seung Il

6. Metal-Ion-Incorporated Fullerene Nanowhiskers and Size-Tunable Nanosheets 75
   Marappan Sathish and Kun'ichi Miyazawa

7. Fabrication and Characterization of C_{60} Fine Crystals and Their Hybridization 89
   Akito Masuhara, Zhenquan Tan, Hitoshi Kasai, Hachiro Nakanishi, and Hidetoshi Oikawa

8. In Situ Transmission Electron Microscopy of Fullerene Nanowhiskers and Related Carbon Nanomaterials 103
   Tokushi Kizuka

9. Mechanical Bend Testing of Fullerene Nanowhiskers 117
   Stepan Lucyszyn and Michael P. Larsson

10. Magnetic Alignment of Fullerene Nanowhiskers 137
    Guangzhe Piao, Fumiko Kimura, and Tsunehisa Kimura

11. Optical Properties of Fullerene Nanowhiskers　　147
    Kiyoto Matsuishi

12. Surface Nanocharacterization of Fullerene Nanowhiskers　　163
    Daisuke Fujita and Mingsheng Xu

13. Structural and Thermodynamic Properties of Fullerene Nanowhiskers　　185
    Hideaki Kitazawa and Kenjiro Hashi

14. High-Temperature Heat Treatment of Fullerene Nanofibers　　197
    Ryoei Kato, Kun'ichi Miyazawa, Toshiyuki Nishimura, Zheng-ming Wang, and Tokushi Kizuka

15. Electronics Device Application of Fullerene Nanowhiskers　　209
    Yuichi Ochiai, Nobuyuki Aoki, and Jonathan Paul Bird

Index　　229

Chapter 1

INTRODUCTION TO FULLERENE NANOWHISKERS

Kun'ichi Miyazawa

Fullerene Engineering Group, National Institute for Materials Science,
1-1, Namiki, Tsukuba, Ibaraki 305-0044, Japan
miyazawa.kunichi@nims.go.jp

Fine fibrous precipitates of C_{60} were discovered in 2001 in a colloidal solution of lead zirconate titanate containing C_{60}. Those fine solid fibers of C_{60} were identified as single-crystal C_{60} nanofibers and named "C_{60} nanowhiskers." The synthetic method of preparing C_{60} nanowhiskers — the liquid–liquid interfacial precipitation method — was soon developed and has been widely used to prepare various fullerene nanowhiskers of C_{70}, C_{60} derivatives, and C_{60}. The liquid–liquid interfacial precipitation method is applicable to preparing tubular fullerene nanofibers (fullerene nanotubes), C_{60} nanosheets, and low-dimensional fullerene nanomaterials containing various metals and chemicals. This introductory chapter gives an overview of the synthetic method, the crystallographic structure, and the mechanical, thermal, electrical, and chemical properties of fullerene nanowhiskers and fullerene nanotubes and their application.

Fullerene Nanowhiskers
Edited by Kun'ichi Miyazawa
Copyright © 2012 Pan Stanford Publishing Pte. Ltd.
www.panstanford.com

1.1 WHAT IS A FULLERENE NANOWHISKER?

Lead zirconate titanate (PZT) is a well-known ferroelectric ceramic material widely used in sensors and actuators. PZT thin films are often used in microelectromechanical systems such as microcantilevers.[1] Since the non-ferroelectric pyrochlore phase of PZT appears at low-temperature calcination, it was necessary to find a method to suppress the formation of the pyrochlore phase.

It was known that the vacuum annealing of pyrolized amorphous PZT gel enhances the growth of the perovskite phase by suppressing the formation of the pyrochlore phase.[2] Hence, C_{60} was expected to act as a scavenger for oxygen and suppress the formation of the pyrochlore phase. In fact, it was found that the temperature at which the perovskite phase is formed can be lowered to 400°C from 600°C when a PZT sol containing C_{60} is used to fabricate PZT thin films.[3]

Furthermore, we noticed fine fibrous precipitates formed in the C_{60}-added PZT sol as shown in Fig. 1.1. These fibrous precipitates were found to be single-crystal nanofibers composed of C_{60} and were named "C_{60} nanowhiskers" (C_{60}NWs).[4]

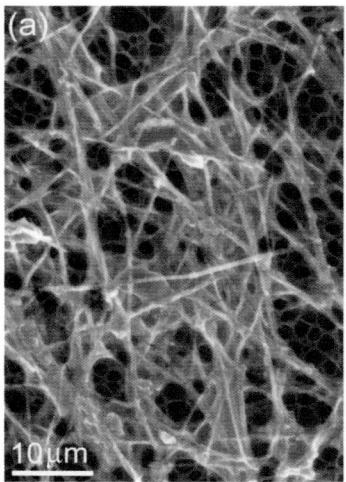 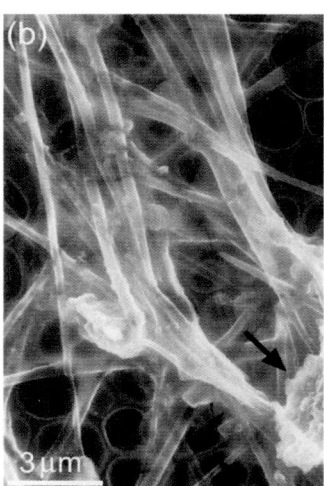

Figure 1.1 (a) Scanning electron microscopy (SEM) image of the C_{60}NWs formed in a PZT sol containing C_{60}. The arrowed part of (b) shows a piece of PZT gel.[4]

The term "C_{60} whisker" was first used by Yosida in 1992.[5] However, the reported C_{60} whisker had a diameter greater than 1 μm and a rugged surface.

A synthetic method for preparing C_{60}NWs with smooth surfaces was soon developed after the discovery of C_{60}NWs. This method was named "liquid–liquid interfacial precipitation method" (LLIP method).[6] Figure 1.2 shows an example of C_{60}NWs synthesized by the LLIP method.

Particles with an aspect ratio (length to diameter) of greater than 3 are defined as fibers.[7] Fullerene nanowhiskers (FNWs) are crystal thin solid fibers that are composed of all species of fullerene molecules, i.e., C_{60}, C_{70}, endohedral fullerenes, fullerene derivative molecules with functional groups, and so forth. FNWs have diameters less than 1000 nm and are usually single crystals.

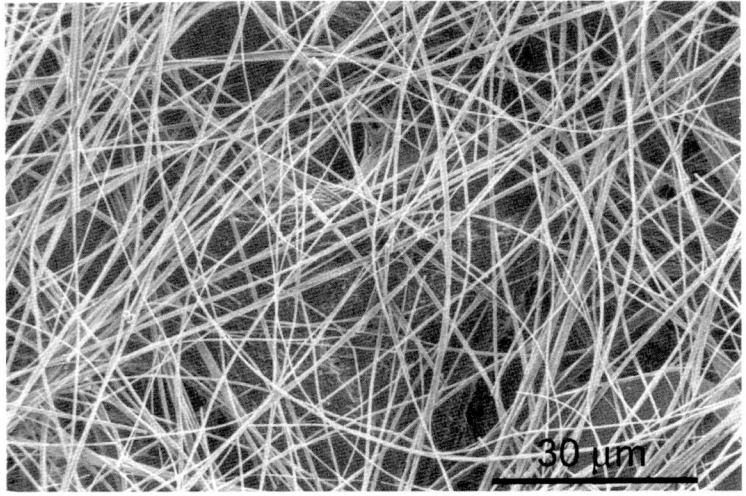

Figure 1.2 SEM image of the C_{60}NWs synthesized by the liquid–liquid interfacial precipitation method.

On the other hand, fullerene nanotubes (FNTs) are thin fibers with tubular morphology and have diameters of less than 1000 nm. FNTs are literally composed of fullerene molecules and can take single-crystal, polycrystalline, or amorphous structures.

FNWs and FNTs can be collectively called "fullerene nanofibers" according to the above definition of fibers. The term "fullerene nanowhisker" is sometimes also called "fullerene nanorod." Although the terms "fullerene nanorod," "fullerene nanowire," "fullerene nanoribbon," "fullerene nanobelt," and so forth are occasionally used, they can all be classified into the category of "fullerene nanofiber."

This introductory chapter reviews the synthetic method and the crystallographic, mechanical, thermal, electrical, and chemical properties of FNWs and FNTs. This chapter also discusses various other forms of fullerene nanomaterials such as fullerene nanosheets.

1.2 THE LLIP METHOD

Fullerene nanofibers (FNFs) with various components and compositions can be synthesized by using the LLIP method, where a poor solvent of fullerene is added to a good solvent solution of fullerene. The mixing order of two liquids can be changed. Alcohols such as ethanol, isopropyl alcohol (IPA), and isobutyl alcohol are often used as the poor solvents, while toluene, m-xylene, benzene, CCl_4, and so forth are used as the good solvents of fullerenes. The liquid–liquid interface formed between a fullerene-saturated good solvent and a poor solvent provides the heterogeneous nucleation sites of fullerene crystals; i.e., the poor solvent plays the role of a nucleation agent.

An example using the LLIP method to synthesize C_{60}NWs is as follows:[6] 5 mL of toluene solution saturated with C_{60} is put in a transparent glass bottle, and 5 mL of IPA is slowly added to the toluene solution to form a liquid–liquid interface. The solution temperature is usually set to be lower than 21°C. Fine C_{60} crystals nucleate at the liquid–liquid interface. The glass bottle is capped and stored for a few minutes or more to grow the C_{60}NWs in an incubator. Figure 1.3 shows a transmission electron microscopy (TEM) image of a C_{60}NW synthesized by using the LLIP method.

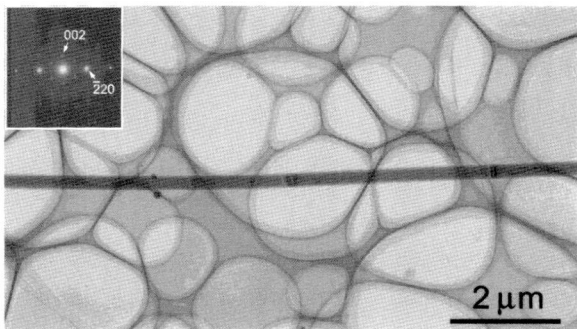

Figure 1.3 TEM image of the first sample of C_{60}NW synthesized by the LLIP method and its SAEDP.[6]

From a high-resolution TEM (HRTEM) image of a C_{60}NW, the intermolecular distance between adjacent C_{60} molecules along the growth axis was found to be 0.962 nm, which is about 4% smaller than that between pristine face-centered cubic (fcc) C_{60} crystals with a lattice constant of a = 1.417 nm. It is conjectured that the shrinkage in the intermolecular distance is caused by the polymerization of C_{60} due to the electron beam irradiation during the TEM observation.[8] A model of polymerized C_{60}NW was first proposed as shown in Fig. 1.4.[6]

Furthermore, it was found that the C_{60}NWs dried in air were composed of the C_{60} molecules bonded via van der Waals forces followed by photopolymerization due to irradiation of laser beams with a wavelength of 532 nm.[9]

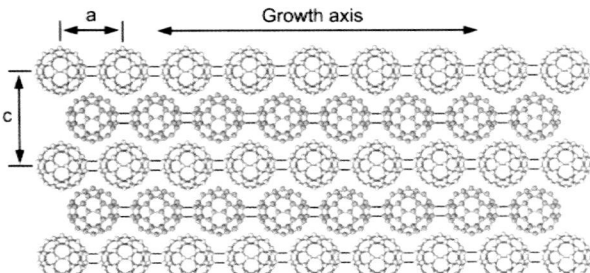

Figure 1.4 Model of one-dimensionally polymerized C_{60}NW with body-centered tetragonal crystal structure with lattice constants of a = 0.962 nm and c = 1.50 nm.[6]

The LLIP method has the following advantages:[10]

1. The synthesis can be performed at room temperature in air using simple apparatuses such as glass bottles, refrigerators, and incubators.
2. Various elements can be doped into FNFs in solution.
3. Multicomponent FNFs composed of different kinds of fullerene molecules can be synthesized.
4. The diameter of FNFs can be controlled by selecting solvent composition, growth temperature, and other factors that influence the nucleation and growth of fullerene embryo crystals.
5. Not only FNFs but also fine powders and nanosheets of fullerene molecules can be obtained by using the LLIP method.[11,12]

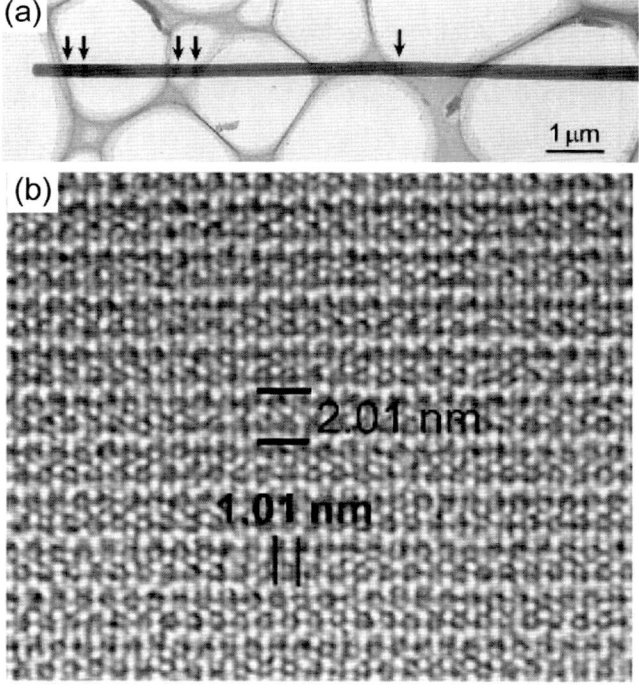

Figure 1.5 (a) TEM image of an iodine-doped single-crystal C_{60}NW with a diameter of 280 nm, (b) HRTEM image of a single-crystal iodine-doped C_{60}NW taken at an underfocusing condition.[13]

Figure 1.5a shows a TEM image of C_{60}NW doped with iodine by using the LLIP method.[13] The dark bands indicated by arrows are the extinction fringes that are formed by the interference of electron beams. The continuous extinction fringes show that the C_{60}NW is a single crystal. The HRTEM image in Fig. 1.5b shows that C_{60} molecules are densely packed along the whisker growth axis. It is noted that the inermolecular distance between C_{60} molecules is 1.01 nm, which is greater by 1% than that between pristine fcc C_{60} crystals with a lattice constant of 1.417 nm. This expansion reflects the incorporation of iodine atoms. The iodine-doped C_{60}NWs exhibit a dislocation-free structure with a very high crystallinity. It is suggested that the doped iodine atoms obstruct the movement of dislocations and stabilize the solvated structure of C_{60}NWs as described later.

Kobayashi *et al.* investigated the influence of light illumination on the growth of C_{60}NWs. They showed that the growth of C_{60}NWs was markedly enhanced by illumination of fluorescent room light. Furthermore, they showed that the growth rate of C_{60}NWs was significantly promoted by exposing the solution to light of wavelengths between 600 and 625 nm.[14]

It has been shown that the growth rate of C_{60}NWs depends on the small amount of water contained in IPA.[15] The cluster size and structure of IPA molecules are conjectured to be changed by the addition of water and influence the growth condition of C_{60}NWs.[16]

FNTs can also be synthesized by using the LLIP method just as FNWs. The first synthesis of FNTs by using the LLIP method was performed in C_{70} nanotubes (C_{70}NTs) as follows:[17] 2 mL of pyridine solution saturated with C_{70} was prepared and put into a transparent glass bottle (A). Then, 6 mL of IPA was gently added to the pyridine solution to form a liquid–liquid interface at a temperature of about 10°C, and the bottle was stored for several days at 10°C in an incubator. As shown in Fig. 1.6, C_{70}NTs exhibit metallic reddish brown color. Figure 1.7a shows a TEM image of a C_{70}NT that was synthesized by the use of a C_{70}-saturated pyridine solution and isobutyl alcohol.[18] C_{60} nanotubes (C_{60}NTs) containing C_{70} molecules were also synthesized after the synthesis of C_{70}NTs, where the composition of the two-component C_{60}–C_{70}NTs was C_{60}–15 mol%C_{70}.[17]

Figure 1.6 Optical micrograph of C_{70}NTs synthesized by use of a C_{70}-saturated pyridine solution and IPA. (Reprinted with permission from Miyazawa et al.,[18] © 2005, IOP Publishing Ltd.)

The LLIP method can be used in the synthesis of nanosheets composed of C_{60} molecules as well.[12] C_{60} nanosheets with various sizes can be prepared by using a modified LLIP method as shown in Chapter 6, using a C_{60}–CCl_4 solution and alcohols (IPA, ethanol, methanol). The diameter of C_{60} nanosheets can be controlled by changing the combination of solvents from 2.5 µm (CCl_4/ethanol) to 500 nm (CCl_4/methanol).

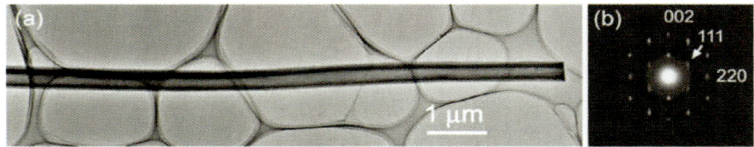

Figure 1.7 (a) TEM image of a C_{70} NT and (b) its SAEDP.[18]

The synthesis of C_{60}NTs (Fig. 1.8) was more difficult than that of C_{70}NTs. However, single-crystal C_{60}NTs can be synthesized by using modified methods described in Refs. 19–23. First, a pyridine solution saturated with C_{60} is prepared. The reddish purple color of the solution gradually turns to brown, which may be attributed to the formation of C_{60}-pyridine adducts.[24] The change in color is enhanced by illumination of ultraviolet (UV)–visible light. Then, 1 mL of the brown pyridine solution of C_{60} is poured into a transparent 10 mL glass bottle and 9 mL of IPA is added. The solution is mixed in an ultrasonic bath for 1 min and is stored in an incubator at 10°C. C_{60}NTs appear on the bottom of the bottle in a few days. The ultrasonication assists the nucleation and dispersion of fullerene embryo crystals.

Figure 1.8 (a) Optical micrograph of as-prepared C_{60}NTs and (b) magnified image of the C_{60}NTs. (c) Cross-sectional TEM image for the C_{60}NTs. (Reprinted with permission from Miyazawa et al.,[23] © 2008, Elsevier.)

The nucleation and growth process of C_{60}NTs was minutely investigated for finding out the effect of solvent ratio, light illumination, and temperature. The yield of C_{60}NTs was found to be dependent on the C_{60}-pyridine solution/IPA ratio, the growth temperature, and the wavelength of light illuminated on the initial pyridine solution saturated with C_{60}. These results indicate that the nucleation of C_{60}NTs is controlled by the degree of C_{60} supersaturation and the size and structure of the clusters of C_{60} and solvents.

The mean diameter of C_{60}NTs also changes depending on the volume ratio of C_{60}-saturated pyridine solution and IPA. Thinner C_{60} fibers were obtained by decreasing the amount of IPA added to the C_{60}-saturated pyridine solution, suggesting that a more concentrated C_{60}-pyridine solution yields C_{60} crystal nuclei with a higher spatial density and smaller sizes.

The yield of C_{60}NTs became the highest when the C_{60}-saturated pyridine solution was illuminated by the light of 370 nm wavelength, which shows the maximum absorption peak of solid C_{60}. However, the yield of C_{60}NTs was shown to be high also in the wavelength range of 600–800 nm, though the light absorption by C_{60} is weak in this wavelength range. This phenomenon is considered to be related to the transient absorption of triplet excited state of C_{60} in the region around 740 nm that is formed by the decay of the photoexcited singlet C_{60} through intersystem crossing.[25]

FNWs can be synthesized by adding fullerene solutions to alcohols as well.[26–28] Recently, vertically aligned fullerene microtubes were successfully fabricated on porous alumina membranes as shown in Fig. 1.9,[29,30] where IPA was slowly injected into C_{60}-saturated toluene solutions through the alumina membranes. The length of the vertically grown C_{60} microtubes reaches about 500 µm, and their planar density and diameter can be modified by changing the synthesis conditions such as the injection rate of IPA and the amount of C_{60}-saturated toluene solution.

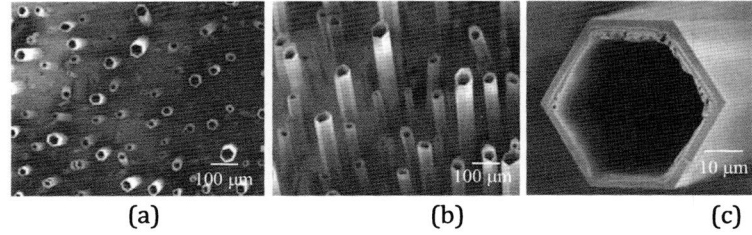

Figure 1.9 SEM images for the vertically aligned C_{60} microtubes. (a) plane view image, (b) tilted image, and (c) enlarged image. (Reprinted with permission from Toita et al.,[30] ©2009, IOP Publishing Ltd.)

As shown above, the LLIP method can be extended variously. Porous C_{60}NWs with a high specific surface area of 376 m²/g were obtained by combining the LLIP method and ultrasonication, using a C_{60}-saturated benzene solution, and IPA.[31] Furthermore, the LLIP method was extended to the synthetic method by using microchannel reactors. The first synthesis of C_{60}NWs with the microchannel reactor system was conducted using a microchannel with a width of 250 µm, a depth of 123 µm, and a length of 22.5 mm.[32] A toluene solution of C_{60} and IPA were introduced through the inlets formed on the device, where C_{60}NWs were synthesized at the liquid–liquid interface of a laminar flow as shown in Fig. 1.10.

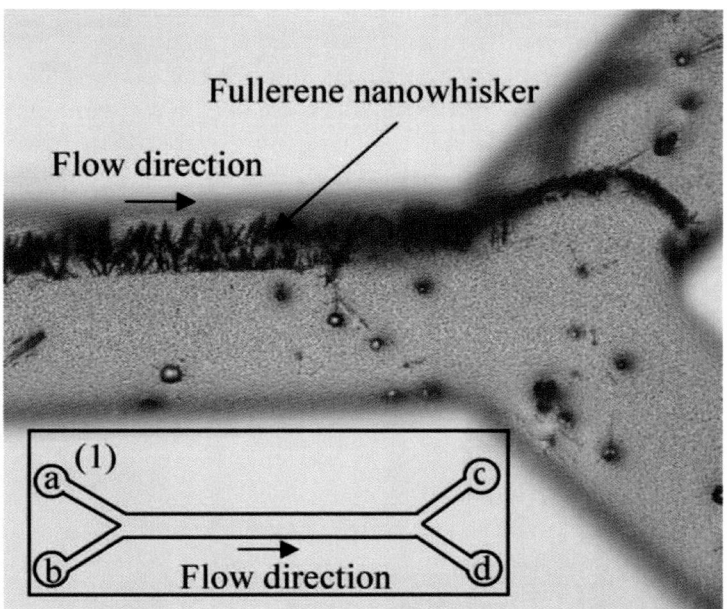

Figure 1.10 Synthesis of C_{60} fullerene nanowhiskers by the microchannel reactor. Reprinted with permission from Lee et al.,[32] © 2004, Elsevier.)

The synthesis of low-dimensional C_{60} nanomaterials was also performed by Shinohara et al. using similar microchannel devices. They produced various forms of C_{60} such as ribbons, tubes, dendrites, spheres, prisms, columns, needles, and plates.[33,34]

1.3 STRUCTURAL CHARACTERIZATION OF FNWs AND FNTs

As-grown FNFs contain solvent molecules in general. For example, the as-grown C_{60}NWs synthesized by the use of a toluene solution of C_{60} and IPA have a hexagonal structure with lattice constants $a = 2.405$ nm and $c = 1.001$ nm,[35] while the as-grown C_{60}NWs prepared by the use of a m-xylene solution of C_{60} and IPA have a hexagonal structure with lattice constants $a = 2.407$ nm and $c = 1.018$ nm, which are slightly larger than those of C_{60}NWs synthesized by the use of toluene and IPA.[35] This result may reflect the larger molecular size of *m*-xylene than toluene. Although the above C_{60}NWs turn to an fcc structure upon drying through the evaporation of solvent molecules, the hexagonal structure of C_{60}NWs synthesized by the use of *m*-xylene and IPA was more stable than that of C_{60}NWs synthesized by the use of toluene and IPA.[35]

As shown in Fig. 1.11, C_{60}NWs have holes with a diameter similar to the size of C_{60} molecules along their growth axis. Holes can incorporate solvent molecules and various other elements such as Ni and Fe.[36,37]

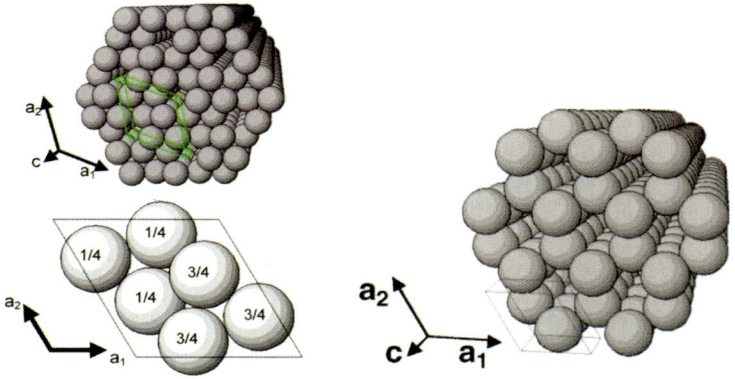

Figure 1.11 (a) Hexagonal structure model of C_{60}NW prepared from toluene solution of C_{60} and IPA. (Reprinted with permission from Minato et al.[35] © 2005, Elsevier.) (b) Hexagonal structure model of the wall of C_{60} nanotube prepared from pyridine solution of C_{60} and IPA. (Reprinted with permission from Minato and Miyazawa,[38] © 2006, Elsevier.)

Like C_{60}NWs, C_{60}NTs synthesized from a pyridine solution of C_{60} and IPA have a hexagonal crystal structure with lattice constants $a = 1.541$ nm and $c = 1.00$ nm.[38] Their hexagonal structure was also transformed into an fcc structure upon drying. The walls of C_{60}NTs also have holes along their growth axis. The holes may incorporate various metal elements as in the case of C_{60}NWs.

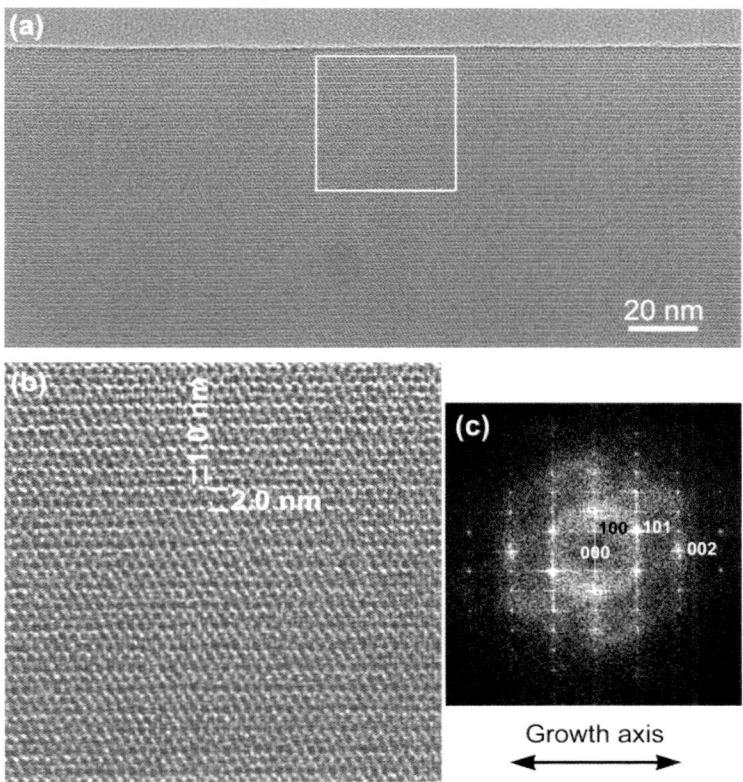

Figure 1.12 (a) HRTEM image, (b) magnified image for the rectangular part of photo (a), and (c) FFT pattern for photo (b) in the C_{60}–4.2 mol% $C_{60}[C(COOC_2H_5)_2]$ nanowhisker. (Reprinted with permission from Miyazawa et al.,[39] © 2006, Institute of Nuclear Chemistry and Technology.)

The structural change in FNFs from the hexagonal phase to the cubic phase upon drying must accompany the reconstruction of crystal lattices through the displacement of C_{60} molecules and the formation of high-density lattice defects such as dislocations and stacking faults. However, as shown in Fig. 4.8 of Chapter 4, hexagonal FNWs with a very good crystallinity were obtained in C_{60}NWs with a composition of C_{60}–4.2 mol% $C_{60}[C(COOC_2H_5)_2]$.[39] C_{60}NWs containing fullerene derivative molecules of $C_{60}[C(COOC_2H_5)_2]$ have an almost constant diameter along the growth axes and are single–crystals as shown in the selected-area electron-diffraction pattern (SAEDP) of the same figure. HRTEM images for the same FNWs are shown in Fig. 1.12. Figure 1.12a shows the smooth whisker surface in atomic scale. The hexagonal lattice parameters a = 2.373 ± 0.031 nm and c = 1.006 ± 0.005 nm were obtained by analyzing Fig. 1.12c and the other HRTEM-FFT (Fast Fourier Transform) patterns. The structure image in Fig. 1.12b shows that the C_{60} cages are densely packed along the whisker growth axis at intervals of 1.0 nm. The above results show that the hexagonal structure of C_{60}NWs can be stabilized by adding appropriate-size solute molecules that can obstruct the migration of dislocations.

C_{70}NTs synthesized by using the pyridine solution of C_{70} and IPA have a solvated hexagonal structure with lattice constants a = 1.603 nm and c = 1.09 nm. However, as in the case of C_{60}NTs, the C_{70}NTs transform to an fcc structure (a = 1.495 ± 0.015 nm) when they lose the contained solvent molecules by drying.[17]

1.4 PHYSICAL AND CHEMICAL PROPERTIES OF FNWs AND FNTs

The mechanical properties of dried C_{60}NWs have been investigated by using a TEM with a function of atomic force microscopy (Kizuka Laboratory at Tsukuba University), and the Young's moduli of C_{60}NWs were measured to range between 32 GPa (160 nm in diameter) and 54 GPa (130 nm in diameter). These values of Young's moduli correspond to 160–650% of C_{60} bulk crystals.[40] The Young's modulus of a C_{60}NT with a diameter of 510 nm was found to be 62–107 GPa by use of the same measurement apparatus.[41] Hence, it is suggested that C_{60}NTs generally have higher Young's modulus than C_{60}NWs owing to the lack of an inner core with the weaker mechanical strength than the surface part. Recently, C_{60}NWs were found to have the core–shell structure with the inner core containing nano-sized pores and the dense surface layer by the cross-sectional observations with TEM.[42]

The plasticity of C_{60}NWs was investigated as follows. Figure 1.13 shows that the surface of C_{60}NWs can be mechanically modified.[43]

A V-groove of 124 nm in width and 70 nm in depth was formed by using a silicon cantilever tip of a scanning probe microscope. The V-groove was formed by scratching the surface of a C_{60} whisker using a Si cantilever tip. The whisker surface was found to be plastically machined in nanometer scale.

It has been known that the electrical resistivity of C_{60} whiskers rapidly decreases with a decrease in their diameter. This unusual phenomenon was first observed in C_{60} whiskers with diameters greater than about 9 μm as shown in Fig. 1.14.[44] Figure 1.14 shows that the resistivity decreases approximately in proportion to the third power of diameter, which means that the unusual resistivity change is due to the bulk effect that may be caused by lattice defects. In reality, high-density dislocations and stacking faults were observed in C_{60}NWs by using TEM.[44,45]

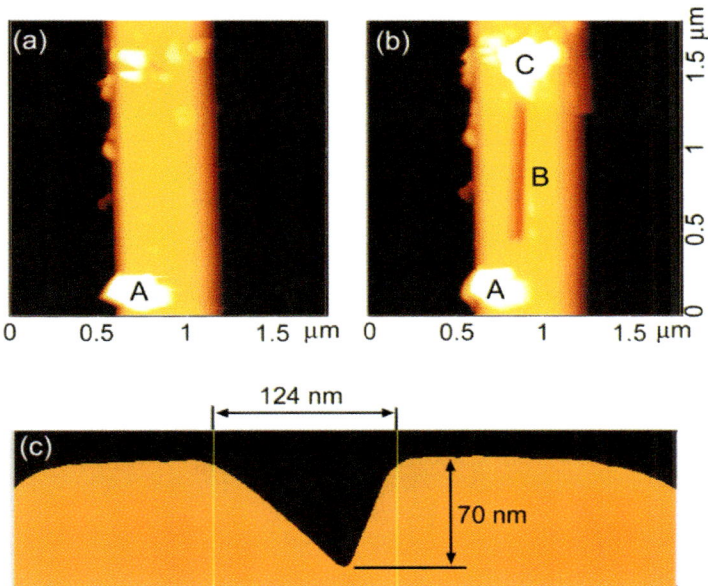

Figure 1.13 SPM images of a C_{60} whisker (SII SPI3800N/SPA-400) (a) before grooving, (b) after grooving with a Si tip, and (c) the cross-sectional profile of the V-groove (b). The marks A and C show the residues formed by the grooving. (Reprinted with permission from Miyazawa et al.,[43] © 2005, SPIE.)

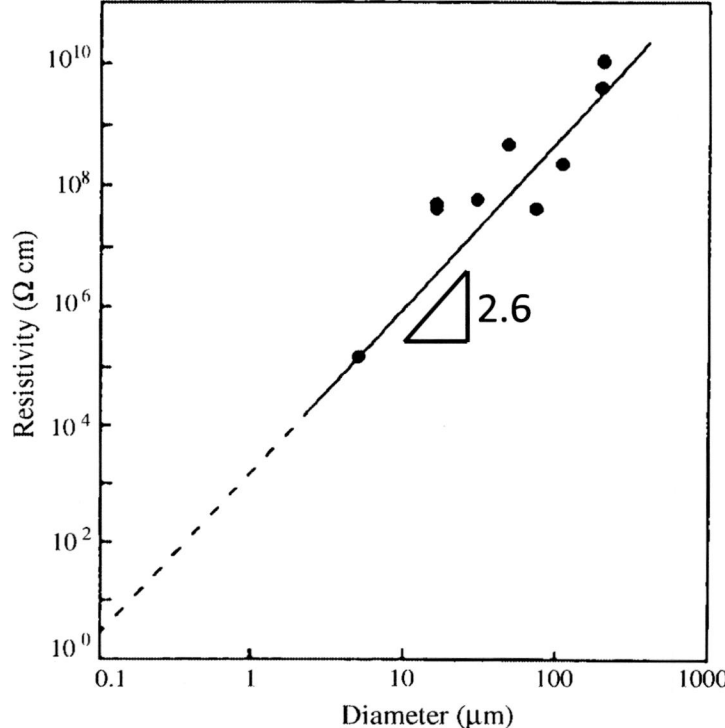

Figure 1.14 Electrical resistivity of C_{60} whiskers measured as a function of diameter. (Reprinted with permission from Miyazawa et al.,[44] © 2003, John Wiley & Sons, Ltd.)

Larsson et al. measured the electrical resistivity of C_{60}NWs with diameters less than 1 μm. A C_{60}NW with a diameter of 650 nm showed a low resistivity of 3 Ω cm.[46] A decrease in lattice defects with decreasing diameter may be one of the reasons for the decrease in resistivity. Impurity elements such as residual solvent molecules and dissolved oxygen atoms may also affect the resistivity change.

The semiconducting nature of C_{60}NWs makes it possible to fabricate field effect transistors (FETs).[47] The first C_{60}NW-FET fabricated by Ogawa et al. exhibited an n-type semiconducting property and a carrier mobility of 2×10^{-2} cm^2/(V s). Recently,

Xu et al. found that C_{60}NWs are covered by thin insulative oxidized layers with a thickness of about 2.5 nm.[48] Hence, the carrier mobility of C_{60}NWs is expected to increase up to the value of 2.0–4.9 cm^2/(V s) of a C_{60} film[49] by removing the insulative oxidized surface layers.

It is an interesting trial to use the insulative surface layer of a C_{60}NW as a dielectric layer of FET, by which an all-carbon FET device may be realized by using a single C_{60}NW.

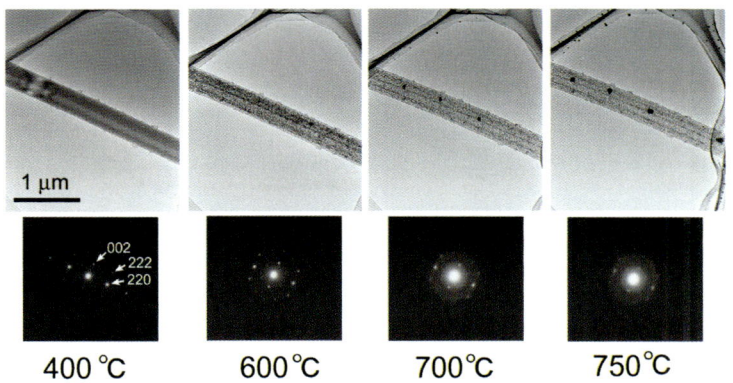

Figure 1.15 In situ observation of a C_{60} nanotube heated in a vacuum of TEM (acceleration voltage 200 kV) at temperatures between 400°C and 750°C. (Reprinted with permission from Miyazawa et al.[20] © 2006, Elsevier.)

In relation to the above electrical properties of C_{60}NWs, Ji et al. reported that C_{60}NWs have a high stability for DC current application and that the fcc C_{60}NWs exhibit higher conductivity than the hexagonal close-packed C_{60}NWs.[28]

C_{60}NTs begin to decompose at 416°C in air and at 708°C in N_2.[50,51] The C_{60}NTs, however, retained their crystalline structure at high temperature in vacuum under irradiation with electron beams.[20] Figure 1.15 shows an in situ TEM observation of a C_{60}NT heated at temperatures between 400 and 750°C for an acceleration voltage of 200 kV.[20] The C_{60}NT turned to a porous structure by heating but managed to retain its tubular structure during the heating experiment. It is also noticed from SAEDPs that C_{60}NTs can maintain their crystalline structure up to 750°C. This high thermal stability of C_{60}NTs in vacuum also must be caused by the polymerization of C_{60} molecules by the electron beam irradiation.[8] It has also been reported that the polymerization of C_{60}NWs occurs by the irradiation of laser beam of Raman spectroscopy apparatus,[9] or by the irradiation of γ-ray.[52] However, Rauwerdink *et al.* reported that C_{60}NTs were not stable in air and their shape was changed by heating at 180°C for a long time.[53]

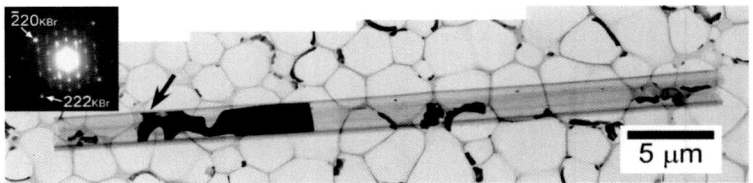

Figure 1.16 TEM image of the KBr crystals deposited inside the C_{60} tube. The SAEDP of the arrowed part shows the single crystalline structure of the KBr crystal.[54]

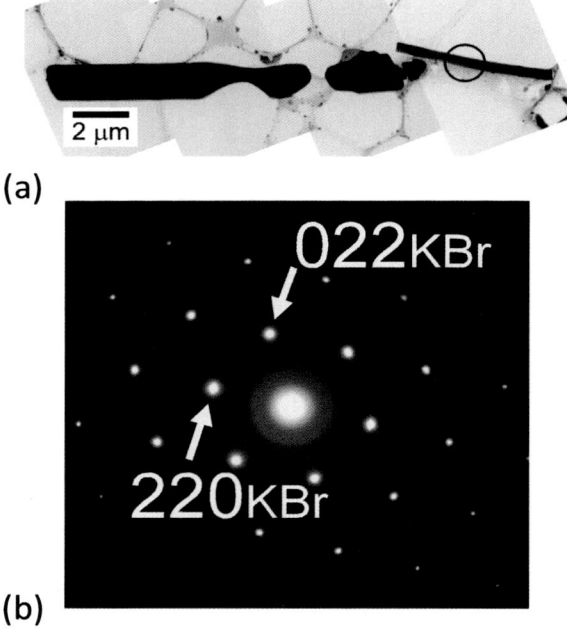

Figure 1.17 (a) TEM image of the extracted KBr crystals and (b) SAEDP of the needlelike KBr crystal marked by the circle of photo (a).[54]

C_{60} nanofibers can be used as templates for chemical synthesis of various nanomaterials. For example, $PtCl_4$ can be deposited into the C_{60}NTs that were cut short by ultrasonication. The materials synthesized inside the C_{60} tubes can be extracted by dissolving their walls by using organic solvents. Single-crystal KBr needlelike crystals were synthesized inside C_{60} tubes by depositing a methyl alcohol solution of KBr using capillary phenomenon and extracted by dissolving the C_{60} tubes with toluene.[54] For example, Fig. 1.16 shows a C_{60}NT containing the deposited KBr crystals inside. The extracted needlelike KBr crystals are shown in Fig. 1.17. The SAEDP in Fig. 1.17b shows that the needlelike KBr crystal is a single crystal. Since the dissolved C_{60} can be used again, the nano- and microtubes of C_{60} have an excellent recyclability.

1.5 SUMMARY

The nanowhiskers, nanotubes, and nanosheets made of fullerene molecules are establishing a new field of low-dimensional carbon nanomaterials. Fullerene nanofibers (fullerene nanowhiskers and fullerene nanotubes) and heat-treated fullerene nanofibers can be applied in transistors and fuel cell electrodes and serve as carriers of various catalysts and chemicals, templates for chemical synthesis, filters, solar cells, and so forth.[55,56] The surface of C_{60}NWs can be machined by using scanning probe microscope tips, and the flexible C_{60}NWs may be utilized as semiconducting light cantilever beams for microelectromechanical systems application. FNFs can be modified by incorporating various metal atoms, chemicals, and fullerene derivatives that add new functions. The chemical bonding properties of FNFs can also be changed by heating or irradiating with light, electron beams, or γ-rays.

In future, the inner and outer surfaces of FNFs will be variously designed to produce a lot of novel functional polymers. The fullerene nanomaterials discussed in this chapter should find wide application in various fields of materials science, engineering, medicine, pharmacy, and so forth.

The other items that could not be covered in this introductory chapter will be discussed in the following chapters.

References

1. T. Kobayashi, R. Kondou, K. Nakamura, M. Ichiki, and R. Maeda, *Jpn. J. Appl. Phys.*, **46**, 7073 (2008).
2. V. S. Tiwari, A. Kumar, V .K. Wadhaman, and D. Pandey, *J. Mater. Res.*, **13**, 2170 (1998).
3. K. Miyazawa, J. Yano, M. Kaga, Y. Ito, K. Ito, and R. Maeda, *Surf. Eng.*, **16**, 239 (2000).
4. K. Miyazawa, A. Obayashi, and M. Kuwabara, *J. Am. Ceram. Soc.*, **84**, 3037 (2001).
5. Y. Yosida, *Jpn. J. Appl. Phys.*, **31**, L505 (1992).
6. K. Miyazawa, Y. Kuwasaki, A. Obayashi, and M. Kuwabara, *J. Mater. Res.*, **17**, 83 (2002).
7. *Nanoscience and Nanotechnologies: Opportunities and Uncertainties*, The Royal Society and The Royal Academy of Engineering, London, 2004, p. 37.
8. M. Nakaya, T. Nakayama, and M. Aono, *Thin Solid Films*, **464–465**, 327 (2004).
9. M. Tachibana, K. Kobayashi, T. Uchida, K. Kojima, M. Tanimura, and K. Miyazawa, *Chem. Phys. Lett.*, **374**, 279 (2003).
10. K. Miyazawa, *J. Nanosci. Nanotechnol.*, **9**, 41 (2009).
11. Y. Takahashi and K. Asai, *J. Japan Inst. Metals*, **68**, 326 (2004) (in Japanese).
12. M. Sathish and K. Miyazawa, *J. Am. Chem. Soc.* **129**, 13816 (2007).
13. K. Miyazawa and K. Hamamoto, *J. Mater. Res.*, **17**, 2205 (2002).
14. K. Kobayashi, M. Tachibana, and K. Kojima, *J. Cryst. Growth*, **274**, 617 (2005).
15. K. Miyazawa, K. Hotta, R. Kato, J. Fujii, and T. Kizuka, In: *Proceedings of the ASME 2009 Conference on Smart Materials, Adaptive Structures and Intelligent Systems*, Oxnard, CA, September 20–24, 2009, SMASIS2009-1445, SMASIS2009
16. J.-H. Guo, Y. Luo, A. Augustsson, S. Kashtanov, J.-E. Rubensson, D. K. Shuh, H. Ågren, and J. Nordgren, *Phys. Rev. Lett.*, **91**, 157401 (2003).
17. K. Miyazawa, J. Minato, T. Yoshii, M. Fujino, and T. Suga, *J. Mater. Res.*, **20**, 688 (2005).

18. K. Miyazawa, J. Minato, T. Yoshii, and T. Suga, *Sci. Technol. Adv. Mater.*, **6**, 388 (2005).
19. J. Minato, K. Miyazawa, and T. Suga, *Sci. Technol. Adv. Mater.*, **6**, 272 (2005).
20. K. Miyazawa, J. Minato, M. Fujino, and T. Suga, *Diam Relat. Mater.*, **15**, 1143 (2006).
21. C. L. Ringor and K. Miyazawa, *Diam. Relat. Mater.*, **17**, 529(2008).
22. C. L. Ringor and K. Miyazawa, *NANO*, **3**, 329 (2008).
23. K. Miyazawa and C. Ringor, *Mater. Lett.*, **62**, 410 (2008).
24. J. Cheng, Y. Fang, Q. Huang, Y. Yan, and X. Li, *Chem. Phys. Lett.*, **330**, 262 (2000).
25. T. Akasaka, Y. Maeda, T. Wakahara, M. Okamura, M. Fujitsuka, O. Ito, K. Kobayashi, S. Nagase, M. Kako, Y. Nakadaira, and E.Horn, *Org. Lett.*, **1**, 1509 (1999).
26. K. Miyazawa, J. Minato, T. Mashino, T. Yoshii, T. Kizuka, R. Kato, M. Tachibana, and T. Suga, In: *Proceedings of the 2nd JSME/ASME International Conference on Materials and Processing*, Seattle, 2005, SMS23.
27. Y. Jin, R. J. Curry, J. Sloan, R. A. Hatton, L. C. Chong, N. Blanchard, V. Stolojan, H. W. Kroto, and S. Ravi P. Silva, *J. Mater. Chem.*, **16**, 3715 (2006).
28. H. Ji, J. Hu, L. Wan, Q. Tang, and W. Hu, *J. Mater. Chem.*, **18**, 328 (2008).
29. S. Cha, K. Miyazawa, and J. Kim, *Chem. Mater.*, **20**, 1667 (2008).
30. S. Toita, K. Miyazawa, K. Hotta, and M. Tachibana, *J. Phys.: Conf. Ser.*, **159**, 12012 (2009).
31. M. Sathish, K. Miyazawa, and T. Sasaki, *Chem. Mater.*, **19**, 2398 (2007).
32. S.-H. Lee, K. Miyazawa, and R. Maeda, *Carbon*, **43**, 855 (2005).
33. K. Shinohara, T. Fukui, H. Abe, N. Sekimura, and K. Okamoto, *Chem. Lett.*, **35**, 1108 (2006).
34. K. Shinohara, T. Fukui, H. Abe, N. Sekimura, and K. Okamoto, *Langmuir*, **22**, 6477 (2006).
35. J. Minato and K. Miyazawa, *Carbon*, **43**, 2837 (2005).
36. M. Sathish, K. Miyazawa, and T. Sasaki, *Diam. Relat. Mater.*, **17**, 571 (2008).

37. M. Sathish and K. Miyazawa, *NANO*, **3**, 409 (2008).
38. J. Minato and K. Miyazawa, *Diam. Relat. Mater.*, **15**, 1151 (2006).
39. K. Miyazawa, J. Minato, T. Mashino, S. Nakamura, M. Fujino, and T. Suga, *NUKLEONIKA*, **51**(Suppl. 1), S41 (2006).
40. K. Asaka, R. Kato, K. Miyazawa, and T. Kizuka, *Appl. Phys. Lett.*, **89**, 071912 (2006).
41. T. Kizuka, K. Saito, and K. Miyazawa, *Diam. Relat. Mater.*, **17**, 972 (2008).
42. R. Kato and K. Miyazawa, In: *Presentation Extended Abstracts of Carbon 2009, The Annual Conference on Carbon*, Biarritz, France, June 14–19, 2009, Topic 11, p. 418.
43. K. Miyazawa, M. Fujino, J. Minato, T. Yoshii, T. Kizuka, and T. Suga, *Proc. SPIE*, **5648**, 224 (2005).
44. K. Miyazawa, Y. Kuwasaki, K. Hamamoto, S. Nagata, A. Obayashi, and M. Kuwabara, *Surf. Interface Anal.*, **35**, 117 (2003).
45. K. Miyazawa, K. Hamamoto, S. Nagata, and T. Suga, *J. Mater. Res.*, **18**, 1096 (2003).
46. M. P. Larsson, J. Kjelstrup-Hansen, and S. Lucyszyn, *ECS Trans.*, **2**, 27 (2007).
47. K. Ogawa, T. Kato, A. Ikegami, H. Tsuji, N. Aoki, and Y. Ochiai, *Appl. Phys. Lett.*, **88**, 112109 (2006).
48. M. S. Xu, Y. Pathak, D. Fujita, C. Ringor, and K. Miyazawa, *Nanotechnology*, **19**, 075712 (2008).
49. K. Itaka, M. Yamashiro, J. Yamaguchi, M. Haemori, S. Yaginuma, Y. Matsumoto, M. Kondo, and H. Koinuma, *Adv. Mater.*, **18**, 1713 (2006).
50. K. Miyazawa, J. Minato, K. Asaka, T. Kizuka, T. Mashino, S. Nakamura, T. Tachibana, and T. Suga, In: *The Papers of Technical Meeting on Physical Sensor*, IEE, Japan, 2007, PHS-07-14 (in Japanese).
51. K. Miyazawa, C. Ringor, K. Nakamura, M. Tachibana, K. Saito, and T. Kizuka, *IEEJ Trans. SM*, **128**, 317 (2008).
52. S. Malik, N. Fujita, P. Mukhopadhyay, Y. Goto, K. Kaneko, T. Ikeda, and S. Shinkai, *J. Mater. Chem.*, **17**, 2454 (2007).
53. K. Rauwerdink, J.-F. Liu, J. Kintigh, and G. P. Miller, *Microsc. Res. Tech.*, **70**, 513 (2007).

54. J. Minato and K. Miyazawa, *J. Mater. Res.*, **21**, 529 (2006).
55. Q. Wang, Y. Zhang, K. Miyazawa, R. Kato, K. Hotta, and T. Wakahara, *J. Phys.: Conf. Ser.*, **159**, 012023 (2009).
56. P. R. Somani, S. P. Somani, and M. Umeno, *Appl. Phys. Lett.*, **91**, 173503 (2007).

Chapter 2

GROWTH, STRUCTURES, AND MECHANICAL PROPERTIES OF C_{60} NANOWHISKERS

Masaru Tachibana

Department of Nanosystem Science, Yokohama City University, 22-2 Seto,
Kanazawa-ku, Yokohama 236-0027, Japan
tachiban@yokohama-cu.ac.jp

C_{60} nanowhiskers (C_{60}NWs) are obtained from the solution growth by using a liquid–liquid interfacial precipitation method. The growth of C_{60}NWs is promoted by light illumination. Typical C_{60}NWs are less than 500 nm in diameter and greater than 100 μm in length. As-grown C_{60}NWs in the solution have a hexagonal solvated structure. The structure is easily changed into a face-centered cubic (fcc) one by the evaporation of solvent molecules in air. The solvated structures exhibit large elastic deformation. Compared with the solvated structures, the fcc structures in air show small elastic deformation, i.e., relative brittleness,. The fcc structures also exhibit semiconductor properties as do intrinsic fcc C_{60} bulk crystals. Such unique shapes and novel properties of C_{60}NWs are very attractive for various applications in the fields of nanoscale devices, energy, and environment. In this chapter, the growth method and structural characteristics for C_{60}NWs are reviewed. In addition, the mechanical properties related to the structures are shown.

Fullerene Nanowhiskers
Edited by Kun'ichi Miyazawa
Copyright © 2012 Pan Stanford Publishing Pte. Ltd.
www.panstanford.com

2.1 INTRODUCTION

The success in efficiently synthesizing C_{60} by Krätschmer *et al.*[1] in 1990 has generated much interest in the growth of this new class of molecular crystals and their physical properties. C_{60} crystals are grown from solution and vapor saturated with C_{60}. In the solution growth, the solvent molecules are often incorporated into the lattice, leading to various crystal structures and shapes, including whiskers and fibers.[2-5] This means that the solution growth can control the crystal structure and shape. On the other hand, the vapor growth can provide solvent-free crystals.[6-11] The solvent-free crystals grown from vapor of pure C_{60} show a face-centered cubic (fcc) structure condensed with a van der Waals interaction. The large crystals with millimeter-size and prominent facets have been grown by using various vapor growth techniques: temperature gradient,[6] double temperature gradient,[7,8] vapor transport by inert gas,[9] periodic oscillation or descending of the growth temperature,[10] and continuous pulling techniques with double temperature gradient.[11,12] They have been widely used for the measurements of intrinsic physical properties of C_{60} crystals.[13]

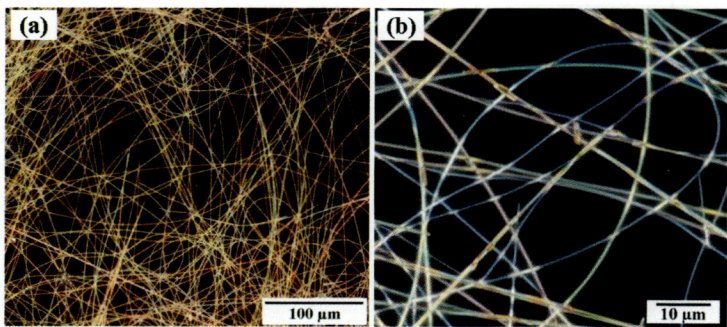

Figure 2.1 Optical micrographs of C_{60}NWs. (a) Low and (b) high magnification.

In 2002, a new type of needlelike C_{60} crystals was obtained from the solution growth by Miyazawa *et al.*[14] When isopropyl alcohol, which is known as a poor solvent of C_{60}, is gently added to a toluene solution saturated with C_{60}, needlelike or fibrous crystals are precipitated at the interface between the solutions. Since their diameter is in the submicron range, these crystals have been referred to as C_{60} nanowhiskers (C_{60}NWs). The preparation method has been referred to as liquid–liquid interfacial precipitation (LLIP) method. The LLIP method has also been successfully used to fabricate fullerene nanowhiskers (FNWs) from higher fullerenes such as C_{70} and C_{60} derivatives.[15–18] These FNWs are also discussed in the review article by Miyazawa.[19] The typical C_{60}NWs are less than 0.5 μm in diameter and greater than 100 μm in length. Moreover, we can easily obtain C_{60}NWs with greater than 1 mm in length or the aspect ratio (the ratio of length to diameter) of greater than 4000. Such a unique shape is quite different from that of C_{60} needlelike crystals reported so far.[20–24] The optical microscopic images of typical C_{60}NWs are shown in Fig. 2.1. Such shapes and structures of FNWs are very attractive for the applications to nanoscale devices. In this chapter, the growth, structure, and mechanical properties of C_{60}NWs are presented.

2.2 GROWTH

2.2.1 LLIP Method

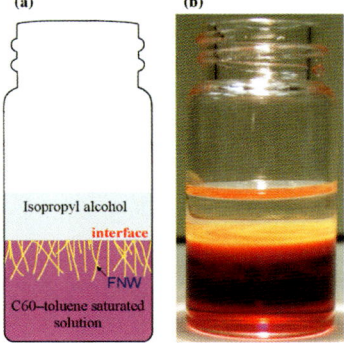

Figure 2.2 LLIP method for the growth of C_{60}NWs. (a) Schematic figure and (b) picture for a growth bottle with a liquid–liquid interface of isopropyl alcohol and toluene solution saturated with C_{60}.

C_{60}NWs can be grown by using the LLIP method. The typical growth procedure is as follows. As-received C_{60} powder is used as a source material. The toluene solution saturated with C_{60} is prepared and poured into a glass bottle. Then, isopropyl alcohol is gently added into the bottle to form a liquid–liquid interface, where the upper part is isopropyl alcohol, which is a poor solvent for C_{60}, and the lower part is the toluene solution saturated with C_{60}. The picture and schematic figure for the growth bottle with the liquid–liquid interface in the above procedure are shown in Fig. 2.2. The bottle is loosely capped and kept at 21°C.

The nucleation of C_{60}NWs occurs at the liquid–liquid interface. The interface disappears due to the diffusion of the solution within 24 h after the formation of the interface. So the period of 24 h including the formation of the interface is called "nucleation period." After the disappearance of the interface, the growth rather than the nucleation mainly occurs. So the period over 24 h after the formation of the interface is called "growth period" After 10 days, C_{60}NWs with 250 nm diameter and greater than 2 mm length are obtained.

Figure 2.3 shows scanning electron micrographs of C_{60}NWs grown by using the LLIP method. Long straight or curved nanowhiskers with constant diameters are individually observed, as shown in Fig. 2.3a. Their lengths reach more than several hundreds of micrometers. The bundled nanowhiskers are also observed. The majority of the nanowhiskers have a diameter of about 250 nm, as seen in Fig. 2.3b. The cross section of each nanowhisker shows a hexagonal shape. As seen in Fig. 2.3c, a starlike cross section with a sixfold symmetry can also be obtained by dissolving the hexagonal nanowhisker in the solution due to the heating. Individual C_{60}NWs are also observed by using a simple optical microscope, as shown in Fig. 2.1a, although the sharpness of the images is worse than that of scanning electron microscopy (SEM) images. They appear yellowish in color. Such an optical microscopic observation means that it is relatively easy to manipulate C_{60}NWs. This can lead to the measurement of physical properties of one isolated nanowhisker.

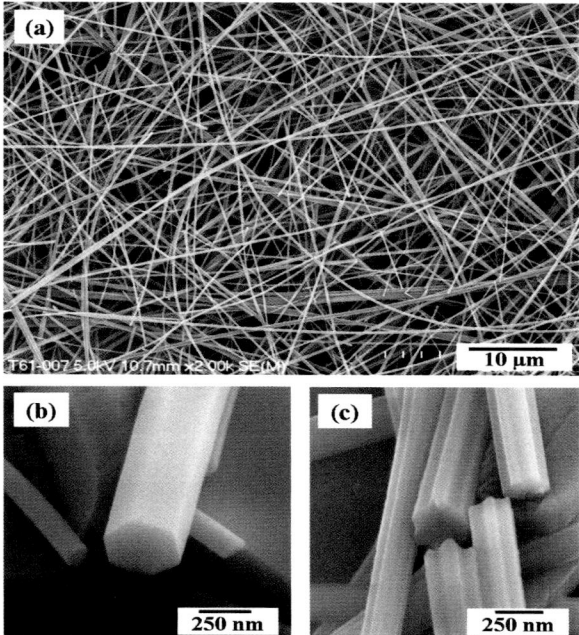

Figure 2.3 SEM images of C_{60}NWs grown by using the LLIP method. (a) Typical SEM images of C_{60}NWs and (b) their cross section with hexagonal shape. (c) Starlike cross section with a sixfold symmetry of C_{60}NWs, which can be obtained by dissolving the hexagonal C_{60}NW in the solution due to the heating. (Reprinted with permission from Tachibana et al.,[25] © 2003, Elsevier, and Kobayashi et al.,[26] © 2004, Elsevier.)

2.2.2 Photo-Assisted Growth

It is well known that C_{60} molecules are very sensitive to light. For example, C_{60} molecules in the solids are easily polymerized under light.[27–30] Such photopolymerization can also occur in the solution.[31] It was found that the growth of C_{60}NWs in the solution is promoted under illumination with weak room light (fluorescent light). The photo-assisted growth of C_{60}NWs strongly depends on the wavelength of light. The illumination wavelength dependence of the number of grown C_{60}NWs is shown in Fig. 2.4. Note that only C_{60}NWs, which are longer than 50 μm and thinner than 500 nm, are counted. As seen in Fig. 2.4, in the wavelength range 365–575 nm, C_{60}NWs are hardly grown. Especially, in C_{60}NWs grown under light with wavelengths below 400 nm, the crystal surface is so rough that no clear crystal habit is observed.

The growth rate of C_{60}NWs is significantly promoted by light illumination with wavelengths only between 600 and 625 nm. The number of the long C_{60}NWs reaches a maximum at 600 nm. No strong wavelength dependence of the diameter of grown C_{60}NWs is observed, and their diameter is about 250 nm.

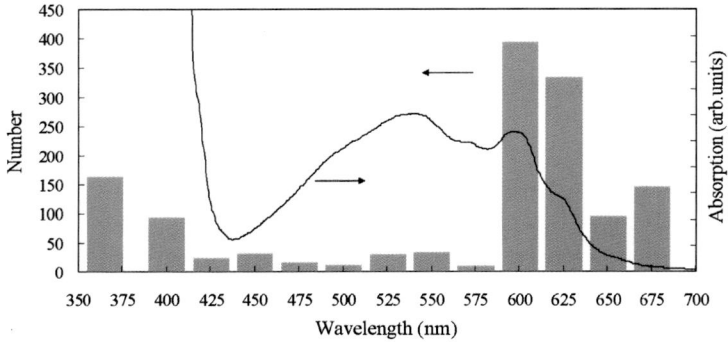

Figure 2.4 Illumination wavelength dependence of the number of grown C_{60}NWs, where only C_{60}NWs, which are longer than 50 μm and thinner than 500 nm, were counted. The measured optical absorption spectrum in the C_{60}-toluene solution is also shown in this figure. (Reprinted with permission from Kobayashi et al.,[26] © 2004, Elsevier.)

To examine the origin of photo-assisted C_{60}NW growth, the optical absorption spectra in the C_{60} toluene solution are also measured. The absorption spectrum in the C_{60} toluene solution is shown in Fig. 2.4. The broad optical absorption is observed over the wavelength range from 450 to 675 nm.[32] Similar optical absorption has also been observed in C_{60} thin films because of the weak intermolecular interaction.[33] The broad optical absorption originates from molecular transitions to lowest excited singlet states in C_{60} molecules.[34] It is known that C_{60} radical anions formed due to electron transfer from polar solvents are formed in quenching processes via the excited singlet states.[35,36] If C_{60} radical anions should predominately contribute to the photo-assisted growth, the number of grown C_{60}NWs is proportional to the relative intensity of optical absorption, indicating the probability of the transition to excited singlet states. However, as shown in Fig. 2.4, C_{60}NWs are hardly grown even under light illumination with 530 nm at which the optical absorption intensity is high. Therefore, it seems that the formation of C_{60} radical anions is not related to the photo-assisted growth.

The photo-assisted C_{60}NW growth is effective in light with only 600 nm. It is known that in the red light around 600 nm also occurs the photo-induced hardening in C_{60} crystals.[37] The red light appears to give rise to a significant illumination effect on both C_{60} solution and solid. According to previous reports,[38,39] the mechanism of the optical absorption around 600 nm is different from that around 530 nm. Therefore, the mechanism of the optical absorption around 600 nm must be related to the photo-assisted growth. The optical absorption around 600 nm originates from the forbidden molecular transition h_u-t_{1u} to lowest excited singlet states in C_{60} molecules.[38,39] These transitions gain nonzero strength through the excitation of an appropriate odd-parity vibration mode (Hertzberg–Teller coupling) and their upper electronic states can be influenced by Jahn–Teller dynamic distortions. Such distortions with the optical absorption can promote the condensation of C_{60} molecules, or the crystallization, in solution. Thus, it is suggested that the formation of excimer-like products by light illumination might accelerate the growth of C_{60}NWs.

2.3 STRUCTURE

2.3.1 X-Ray Diffraction

The LLIP method included some C_{60} bulk crystals as impurities. They predominantly contribute to X-ray diffraction (XRD) and Raman spectra. To clarify the structure of intrinsic C_{60}NWs, purified C_{60}NWs need to be used for XRD and Raman measurements.

Figure 2.5 shows the time evolution of the XRD pattern of purified C_{60}NWs during drying. The XRD pattern taken for 25 min after the sampling from the glass bottle shows sharp peaks.

These peaks are indexed by the hexagonal system with cell dimensions a = 2.43 nm and c = 1.01 nm (a/c = 2.41). The hexagonal peaks gradually become smaller, and then relatively broad peaks appear. The hexagonal peaks are still observed even after 230 min. However, the peaks disappear after vacuum-drying. On the other hand, the broad peaks are observed at positions similar to those for intrinsic C_{60} single crystals and indexed by an fcc system with a cell dimension of a = 1.44 nm. This means that the hexagonal solvated structure is changed into the fcc structure by the evaporation of solvent molecules. The detailed structural change is discussed by Minato and Miyazawa.[41]

A possible crystal structure suggested by XRD results for the solvated structure of C_{60}NWs is drawn in Fig. 2.6. In the figure, the fullerene molecules are represented as the spheres contacting each other. The large channels along the c axis that can contain solvent molecules lie between C_{60} molecules.

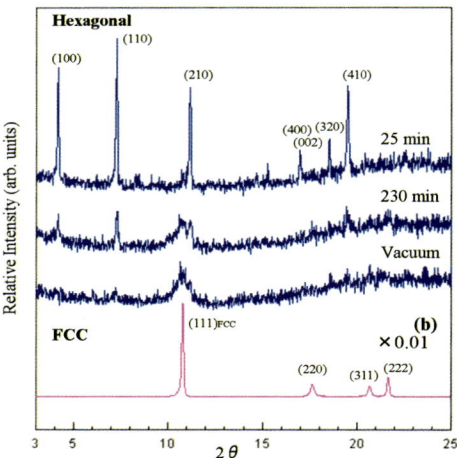

Figure 2.5 Time evolution of the XRD pattern for purified C_{60}NWs during air-drying. The XRD pattern for C_{60}NWs after vacuum-drying are also shown in the figure. For comparison, XRD pattern for intrinsic C_{60} crystals with an fcc structure grown by using the sublimation method is included in the figure. These XRD patterns were obtained with 0.1542 nm of CuKα. (Reprinted with permission from Watanabe et al.,[40] © 2008, The Institute of Electrical Engineers of Japan.)

The packing structure may have symmetry as high as $P6_3/m$ if the positions of the solvent molecules are not taken into account. The structure resembles to that reported by Ramm et al. for bulk crystals obtained through a slow evaporation of *m*-xylene and a mixture of carbon disulfide/*p*-/*m*-xylene solutions of C_{60} (space group $P6_3$, a = 2.37 nm, and c = 1.00 nm at 100 K).[44]

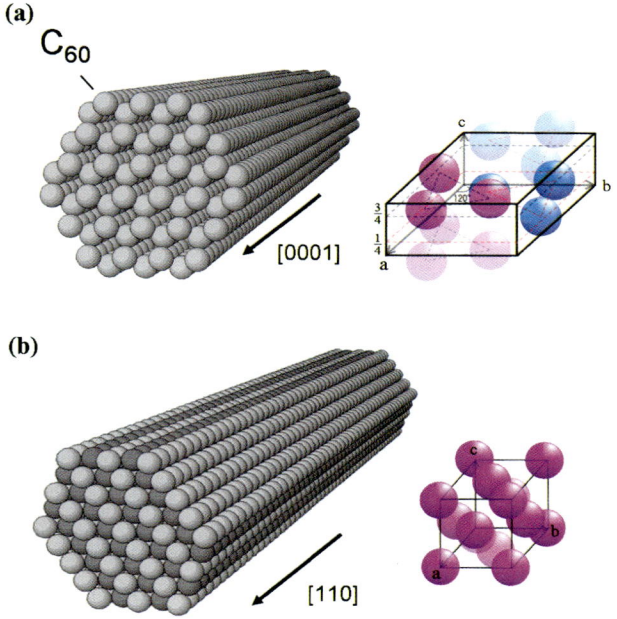

Figure 2.6 Schematic figures of (a) hexagonal solvated structure of C_{60}NWs in solution and (b) fcc structure in air.[42,43]

2.3.2 Raman Spectroscopy

Similar to XRD measurements, the time evolution of the Raman spectrum of purified C_{60}NWs during drying is measured to examine the intermolecular interaction in C_{60}NWs in the solution and air. Figure 2.7 shows the time evolution of main Raman bands at 270, 495, and 1468 cm^{-1} in air, which can be associated with $H_g(1)$ squashing,

$A_g(1)$ breathing, $A_g(2)$ pentagonal pinch modes of C_{60} molecules, respectively.[45] For comparison, Raman bands in C_{60} molecules in a toluene solution, hexagonal C_{60} bulk crystals grown from a toluene solution, and fcc C_{60} bulk crystals grown from sublimation are included in Fig. 2.7.

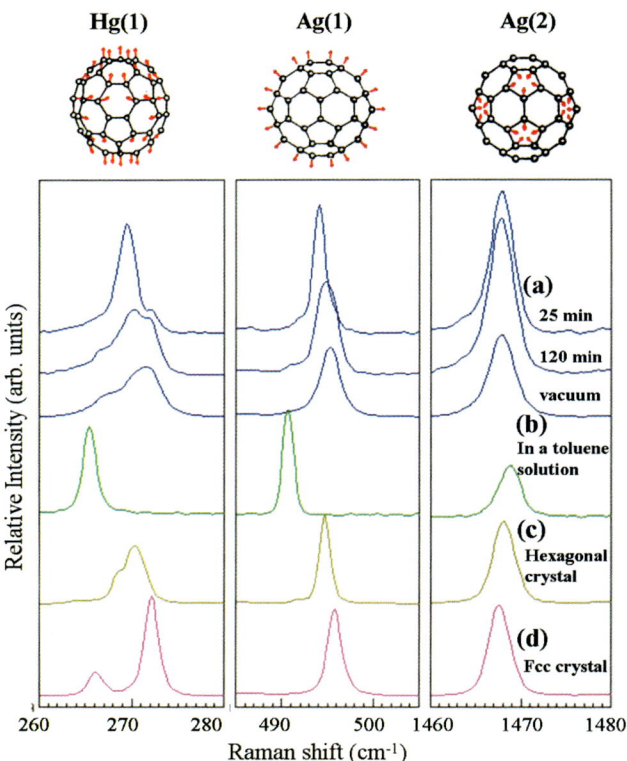

Figure 2.7 Time evolution of main Raman bands for (a) C_{60}NWs during air-drying. The Raman bands at 270, 495, and 1468 cm^{-1} can be associated with $H_g(1)$ "squashing," $A_g(1)$ "breathing," and $A_g(2)$ "pentagonal pinch" modes of C_{60} molecules, respectively. For comparison, Raman bands in (b) C_{60} molecules in toluene solution, (c) hexagonal C_{60} bulk crystals grown from toluene solution, and (d) fcc C_{60} bulk crystals grown from sublimation are included in the figure. The Raman spectra were taken with 1064 nm excitation. (Reprinted with permission from Watanabe et al.,[40] © 2008, The Institute of Electrical Engineers of Japan.)

No significant change in the peak profiles of $A_g(1)$ and $A_g(2)$ modes is observed, although the frequencies of the peaks are shifted. On the other hand, the peak profile of $H_g(1)$ for C_{60} molecules is quite different from that for fcc C_{60} bulk crystals grown from sublimation. Namely, the peak for C_{60} molecules is fitted by a Lorentzian curve while that for the crystals can be fitted by several curves. Such splitting in the $H_g(1)$ mode for the crystals should be attributed to the crystal field due to regular molecular arrangement in the crystals.[45] C_{60}NWs in the solution have the hexagonal structure, which is similar to that of C_{60} bulk crystals grown from a toluene solution. However, no clear splitting in the $H_g(1)$ peak is observed for C_{60}NWs in the solution. This means that C_{60}NWs in the solution have a smaller crystal field, i.e., smaller intermolecular interaction, although they have a regular molecular arrangement, similar to that of C_{60} bulk crystals. It is expected that such an intermolecular interaction in C_{60}NWs with a solvated structure can lead to unique physical properties differing from those in air.

2.4 MECHANICAL PROPERTIES

The mechanical properties of C_{60}NWs in the solution and air are investigated by using a bending method. To observe the bending behavior in the solution, the specimens are mounted on a glass plate with a droplet of the growth solution. The bending deformation for both specimens in the solution and air is carried out by applying a wind pressure with a fan. The bending behavior is in situ observed by using an optical microscope with a charge coupled device.

Figure 2.8 shows the successive pictures of the bending behavior of a C_{60}NW, which is 500 nm in diameter and greater than 100 µm in length. The straight C_{60}NW before applying a wind pressure is indicated by the arrow in Fig. 2.8a. The C_{60}NW is bent with a wind pressure and reaches a curvature of 290 µm, as shown in Fig. 2.8b. Freeing from a wind pressure, C_{60}NW recovers to the original straight shape, as shown in Fig. 2.8c. Such a behavior of the bending shows that C_{60}NWs in the solution exhibit large elastic deformation.

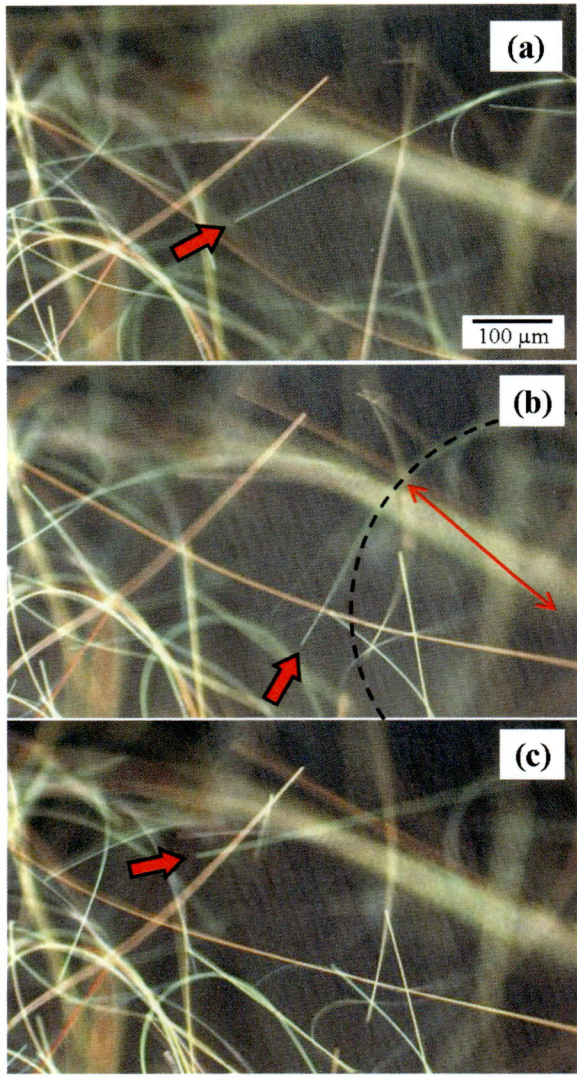

Figure 2.8 Successive pictures of the bending deformation of a C_{60}NW in solution by wind pressure. (a) Straight C_{60}NW before applying wind pressure, (b) bended C_{60}NW with the curvature of 290 μm by wind pressure, and (c) straight C_{60}NW recovered after freeing from the wind pressure. (Reprinted with permission from Watanabe et al.,[40] © 2008, The Institute of Electrical Engineers of Japan.)

The bending experiments are also carried out for C_{60}NWs in air, as shown in Fig. 2.9. A straight C_{60}NW before applying a wind pressure is indicated by the arrow in Fig. 2.9a. The C_{60}NW is bent under a wind pressure, as shown in Fig. 2.9(b). However, the C_{60}NW in air is broken just after it is bent with a curvature of only 650 µm, as shown in Fig. 2.9b. Therefore, C_{60}NWs in air exhibit smaller elastic deformation, i.e., relative brittleness, compared with that in the solution. Thus, C_{60}NWs in the solution show large elastic deformation. This might be due to the solvated structure of C_{60}NWs with a weak intermolecular interaction.

The yield stress or breaking strength for whiskers can be estimated from looplike deformation.[46] In the deformation, the curvature of the loop increases with stress, and further increases in stress leads to yielding or breaking of the whisker. The yield stress σ of the whisker can be estimated from the curvature just before yielding and is given by $\sigma = E\,(r/\rho)$, where E is the Young's modulus of whisker, r is the radius of whisker, and ρ is the curvature of the loop just before yielding.[46] This equation can be obtained from the theory of beam deformation, with $r \ll \rho$. Therefore, the equation is available for the C_{60}NWs with the diameter of 500 nm and the curvature of 650 µm just before breaking in air (Fig. 2.9b). In addition, the Young's modulus of C_{60}NWs in air can be assumed to be 20 GPa corresponding to that of C_{60} bulk crystals,[47,48] since C_{60}NWs in air have the fcc structure condensed with a weak van der Waals interaction similar to that in C_{60} bulk crystals as mentioned above. Substituting these values into the equation mentioned above, the yield stress or breaking strength for C_{60}NWs in air is estimated to be 7.7 MPa. This value is an order as large as 0.7 MPa in C_{60} bulk crystals.[49]

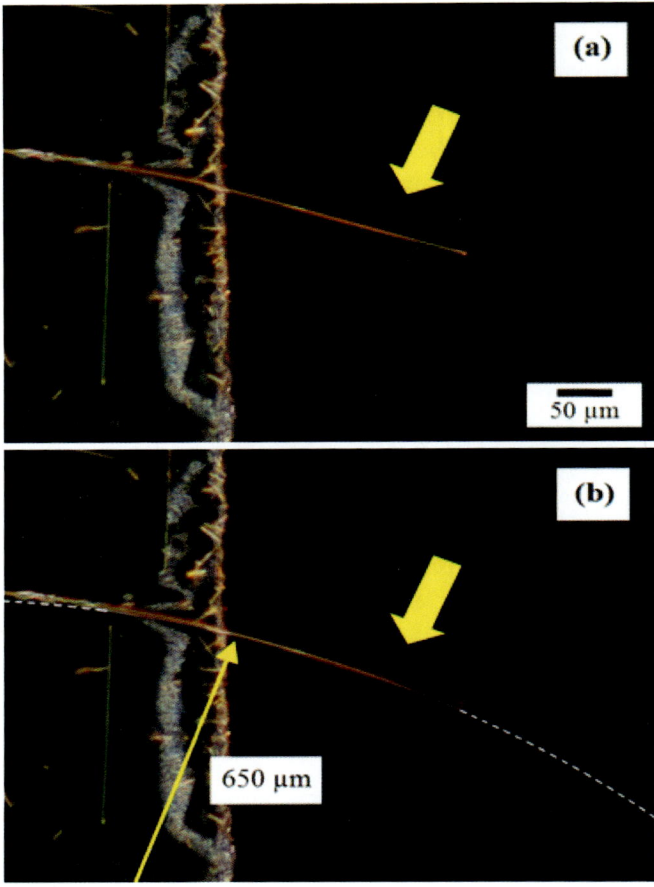

Figure 2.9 Successive pictures of the bending deformation of a C_{60}NW in air. The stress was applied by wind pressure. (a) Straight C_{60}NW before applying wind pressure and (b) bended C_{60}NW with the curvature of 650 μm by wind pressure. The further bending leads to the breaking of C_{60}NW. (Reprinted with permission from Watanabe et al.,[40] © 2008, The Institute of Electrical Engineers of Japan.)

Recently, it was found that the modified C_{60}NWs kept in the growth solution for 3 years retained the hexagonal solvated structure even in air or vacuum.[50] Moreover, the modified C_{60}NWs exhibit higher strength and larger elastic deformation. Such mechanical toughness in air might lead to more potential applications.

2.5 CONCLUSIONS

The LLIP method for the solution growth of C_{60}NWs has been shown. The LLIP method is also available for various FNWs from higher fullerenes such as C_{70} and C_{60} derivatives. The growth of C_{60}NWs using the LLIP method is promoted by light illumination. Especially, the light with 600 nm is effective for the photo-assisted growth. As-grown C_{60}NWs in the solution have a hexagonal solvated structure. The structure is easily changed into an fcc structure by the evaporation of solvent molecules in air. The fcc C_{60}NWs in air exhibit small elastic deformation, i.e., relative brittleness, compared with the hexagonal solvated C_{60}NWs in the solution. In such solvated crystals, the solvent can control not only the growth and shape but also the physical properties. The selection and control of solvents for the solution-grown crystals can be increasingly required for the synthesis of new types of crystalline solids for nanoscale devices.

Acknowledgments

The author is grateful to Mr. Ken-ichi Kobayashi, Mr. Kohei Nakamura, and Ms. Mami Watanabe for many experiments in this chapter. The author thank Mr. Takahiro Yamamoto for preparing figures in this chapter. The author also thank Dr. Kun'ichi Miyazawa and Prof. Kenichi Kojima for fruitful discussions. This work was supported by a Strategic Research Project (K19044) in Yokohama City University.

References

1. W. Krätschmer, L. D. Lamb, K. Fostiropoulos, and D. R. Huffman, *Nature*, **347**, 354 (1990).
2. J. M. Hawkins, S. Loren, A. Meyer, and R. Nunlist, *J. Am. Chem. Soc.*, **113**, 7770 (1991).
3. K. Kikuchi, S. Suzuki, K. Saito, H. Shiramaru, I. Ikemoto, Y. Achiba, A. A. Zakhidov, A. Ugawa, K. Imaeda, H. Inokuchi, and K. Yakushi, *Physica C*, **185–189**, 415 (1991).
4. Y. Yosida, *Jpn. J. Appl. Phys.*, **31**, L505 (1992).

5. A. R. Kortan, N. Kopylov, and F. A. Thiel, *J. Phys. Chem. Solids*, **53**, 1683 (1992).
6. R. L. Meng, D. Ramirez, X. Jiang, P. C. Chow, C. Diaz, K. Matsuishi, S. C. Moss, P. H. Hor, and C.W. Chu, *Appl. Phys. Lett.*, **59**, 3402 (1991).
7. M. Tachibana, M. Michyama, K. Kikuchi, Y. Achiba, and K. Kojima, *Phys. Rev. B*, **49**, 14945 (1994).
8. M. Haluska, H. Kuzmany, M. Vybornnov, P. Rogl, and P. Fejdi, *Appl. Phys. A*, **56**, 161 (1993).
9. M. A. Verheijen, H. Meekes, G. Meijer, E. Raas, and P. Bennema, *Chem. Phys. Lett.*, **19**, 339 (1992).
10. J. Li, S. Komiya, T. Tamura, C. Nagasaki, J. Kihara, K. Kishino, and K. Kitazawa, *Physica C*, **195**, 205 (1992).
11. M. Tachibana, M. Michiyama, H. Sakuma, K. Kikuchi, Y. Achiba, and K. Kojima, *J. Cryst. Growth*, **166**, 883 (1996).
12. K. Kojima, M. Tachibana, Y. Maekawa, H. Sakuma, M. Michiyama, K. Kikuchi, and Y. Achiba, *Crystal Growth of Organic Materials* (Conference Proceedings Series) (American Chemical Society, Washington, DC, 1995), p. 231.
13. M. S. Dresselhaus, G. Dresselhaus, and P.C. Eklund, *Science of Fullerenes and Carbon Nanotubes* (Academic Press, New York, 1996).
14. K. Miyazawa, Y. Kuwasaki, A. Obayashi, and M. Kuwabara, *J. Mater. Res.*, **17**, 83 (2002).
15. K. Miyazawa, *J. Am. Ceram. Soc.*, **85**, 1297 (2002).
16. K. Miyazawa, K. Hamamoto, S. Nagata, and T. Suga, *J. Mater. Res.*, **18**, 1096 (2003).
17. K. Miyazawa, T. Mashino, and T. Suga, *J. Mater. Res.*, **18**, 2730 (2003).
18. K. Miyazawa, T. Mashino, and T. Suga, *Trans. Mater. Res. Jpn.*, **29**, 537 (2004).
19. K. Miyazawa, *J. Nanosci. Nanotechnol.*, **9**, 41 (2009).
20. R. M. Fleming, A. R. Kortan, B. Hessen, T. Siegriest, F. A. Thiel, P. Marsh, R. C. Haddon, R. Tycko, G. Dabbagh, M. L. Kaplan, and A. M. Mujsce, *Phys. Rev. B*, **44**, 888 (1991).
21. H. Moriyama, H. Kobayashi, A. Kobayashi, and T. Watanabe, *Chem. Phys. Lett.*, **238**, 116 (1995).

22. R. Ceolin, J. L. Tamarit, D. O. Lopez, M. Barrio, V. Agafonov, H. Allouchi, F. Moussa, and H. Szwarc, *Chem. Phys. Lett.*, **314**, 21 (1999).
23. S. Ogawa, H. Furusawa, T. Watanabe, and H. Yamamoto, *J. Phys. Chem. Solids*, **61**, 1047 (2000).
24. J. Geng, W. Zhou, P. Skelton, W. Yue, I. A. Kinloch, A. H. Windle, and B. F. G. Johnson, *J. Am. Chem. Soc.*, **130**, 2527 (2008).
25. M. Tachibana, K. Kobayashi, T. Uchida, K. Kojima, K. Tanimura, and K. Miyazawa, *Chem. Phys. Lett.*, **374**, 279 (2003).
26. K. Kobayashi, M. Tachibana, and K. Kojima, *J. Cryst. Growth*, **274**, 617 (2005).
27. A. M. Rao, P. Zhou, K.-A. Wang, G. T. Hager, J. M. Holden, Y. Wang, W.-T. Lee, X.-X. Bi, P. C. Eklund, D. S. Cornett, M. A. Duncan, and I. J. Amster, *Science*, **259**, 955 (1993).
28. M. Matus and H. Kuzmany, *Appl. Phys. A*, **56**, 241 (1993).
29. K. Matsuishi, T. Ohno, N. Yasuda, T. Nakanishi, S. Onari, and T. Arai, *J. Phys. Chem. Solids*, **58**, 1747 (1997).
30. M. Tachibana, H. Sakuma, and K. Kojima, *J. Appl. Phys.*, **82**, 4253 (1997).
31. G. Chambers and H. J. Byrne, *Chem. Phys. Lett.*, **302**, 307 (1999).
32. S. Leach, M. Vervloet, A. Despres, E. Breheret, J. P. Hare, T. J. Dennis, H. W. Kroto, R. Taylor, and D. R. M. Walton, *Chem. Phys.*, **160**, 451 (1992).
33. Y. Wang, J. M. Holden, A. M. Rao, P. C. Eklund, U. D. Venkateswaran, D. E. Eastwood, R. L. Lidberg, G. Dresselhaus, and M. S. Dresselhaus, *Phys. Rev. B*, **51**, 4547 (1995).
34. F. Negri, G. Orlandi, and F. Zerbetto, *J. Chem. Phys.*, **97**, 6496 (1992).
35. T. Nojiri and A. Watanabe, and O. Ito, *J. Phys. Chem. A*, **102**, 5215 (1998).
36. A. S. D. Sandanayaka, Y. Araki, C. Luo, M. Fujitsuka, and O. Ito, *Bull. Chem. Soc. Jpn.*, **77**, 1313 (2004).
37. M. Tachibana, K. Kojima, H. Sakuma, T. Komatsu, and T. Sunakawa, *J. Appl. Phys.*, **84**, 1944 (1998).
38. F. Negri, G. Orlandi, and F. Zerbetto, *J. Phys. Chem.*, **100**, 10849 (1996).

39. J. Hora, P. Panek, K. Navratil, B. Handlirova, J. Humlicek, H. Sitter, and D. D. Stifer, *Phys. Rev. B*, **54**, 5106 (1996).
40. M. Watanabe, K. Miyazawa, K. Kojima, and M. Tachibana, *IEEJ Trans. SM* **128**, 321(2008).
41. J. Minato and K. Miyazawa, *Carbon*, **43**, 2837 (2005).
42. J. Minato, K. Miyazawa, and T. Suga, *Sci. Technol. Adv. Mater.*, **6**, 272 (2005).
43. J. Minato and K. Miyazawa, *Diam. Relat. Mater.*, **15**, 1151 (2006).
44. M. Ramm, P. Luger, D. Zobel, W. Duczek, and J.C.A. Boeyens, *Cryst. Res. Technol.*, **31**, 43 (1996).
45. J. Menéndez and J.B. Page, in *Vibrational Spectroscopy of C_{60}*, eds M. Cardona and G. Güntherodt (Springer, Berlin, 2000).
46. T. Kaneko, *Whiskers* (Kyouritsu Shuppan, Tokyo, 1993).
47. X. D. Shi, A. R. Kortan, J. M. Williams, A. M. Kini, B. M. Saval, and P. M. Chaikin, *Phys. Rev. Lett.*, **68**, 827 (1992).
48. S. Hoen, N. G. Chopra, X.-D. Xiang, R. Mostovoy, Jianguo Hou, W. A. Vareka, and A. Zettl, *Phys. Rev. B*, **46**, 12737 (1992).
49. T. Komatsu, M. Tachibana, and K. Kojima, *Phil. Mag.*, **81**, 659 (2001).
50. M. Watanabe, T. Yamamoto, K. Kojima, M. Tanimura, and M. Tachibana, *J. Phys.: Conf. Ser.*, **159**, 012009 (2009).

Chapter 3

INVESTIGATION OF THE GROWTH MECHANISM OF C_{60} FULLERENE NANOWHISKERS

Kayoko Hotta and Kun'ichi Miyazawa

Fullerene Engineering Group, National Institute for Materials Science, 1-1, Namiki, Tsukuba, Ibaraki 305-0044, Japan
miyazawa.kunichi@nims.go.jp

The growth of C_{60} nanowhiskers (C_{60}NWs) synthesized in a solution was investigated. In order to know the growth mechanism, the length of C_{60}NWs was measured under various growth conditions of temperature, volume mixing ratio of the C_{60}-saturated toluene solution and isopropyl alcohol (IPA), and water content in IPA. A high activation energy of the crystal growth was obtained from the length growth measurement of C_{60}NWs. The growth of C_{60}NWs was found to be enhanced by a small amount of water contained in IPA. It is inferred that the water contained in IPA catalyzes the growth of C_{60}NWs.

Fullerene Nanowhiskers
Edited by Kun'ichi Miyazawa
Copyright © 2012 Pan Stanford Publishing Pte. Ltd.
www.panstanford.com

3.1 INTRODUCTION

C_{60} fullerene nanowhiskers (FNWs) were discovered in 2001 by Miyazawa et al.,[1] and a method to prepare C_{60}NWs was reported in 2002[2] when various researches on the synthesis of FNWs had started. Using the liquid–liquid interfacial precipitation method, various kinds of nanofibers and nanosheets composed of fullerene molecules have been prepared. Although various property characterizations of FNWs have been conducted in order to know their potential application in future, the growth mechanism of FNWs has not been sufficiently cleared. Here the growth of C_{60}NWs under various conditions of the solvent mixing ratio, temperature, and water content in isopropyl alcohol is investigated, and the growth mechanism of C_{60}NWs is discussed on the basis of the length measurement of C_{60}NWs using optical microscopy and scanning electron microscopy.

3.2 EXPERIMENTAL

In the preparation of C_{60}NWs, toluene (99.5% purity) was used as a good solvent of C_{60} and isopropyl alcohol (IPA; 99.7% purity) was used as a poor solvent of C_{60}; 2.8 mg/mL of ground C_{60} powder (99.5% purity) was dissolved in toluene by ultrasonication for 30 min and then filtered to remove the undissolved C_{60} powder. The C_{60}-saturated toluene solution was poured into a 10 mL transparent glass bottle and IPA was slowly added to that solution to form a liquid–liquid interface in a water bath at appropriate temperatures.

In the original liquid–liquid interfacial precipitation method,[2] the bottles were placed in an incubator without mixing, where long C_{60}NWs appeared with some thick needlelike crystals and granular bulk crystals. In this study, however, the bottled solutions were manually mixed after forming the liquid–liquid interface and kept in an incubator at temperatures between 5 and 20°C. In this procedure, short C_{60}NWs were synthesized and almost no granular crystals were formed as shown below.

3.3 TEMPERATURE EFFECT ON THE GROWTH OF C_{60}NWs

The temperatures for the synthesis of C_{60}NWs were set to 5, 10, 15, and 20°C. The length of C_{60}NWs was measured as a function of growth time. C_{60}NWs grow fast in their initial growth stage, and their growth rate decreases with increasing growth time (Fig. 3.1). The length growth rate of C_{60}NWs increased with increasing liquid temperature from 5 to 20°C. To investigate the influence of C_{60} diffusion on the growth of C_{60}NWs, the solutions in the bottles were agitated by magnetic bars at different temperatures. As a result, no significant difference was observed in the average length of C_{60}NWs for each growth temperature and growth time (Fig. 3.2). This result shows that the growth of C_{60}NWs is determined by the surface accumulation process of solute C_{60} molecules.

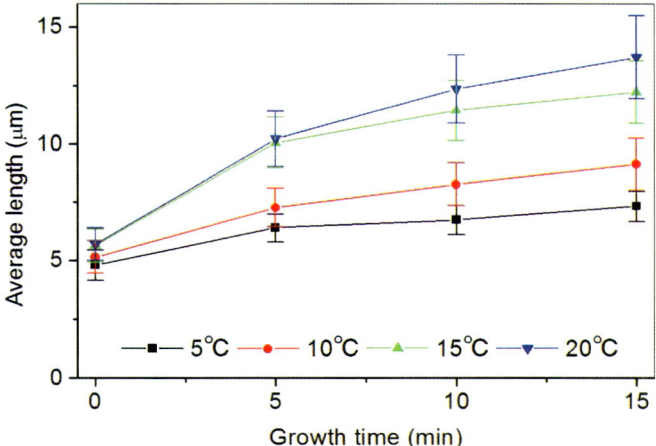

Figure 3.1 Length of C_{60}NWs measured as a function of growth time. Each point shows the average length for 300 C_{60}NWs. (Reprinted with permission from Hotta,[3] © 2008, World Scientific Publishing Company.)

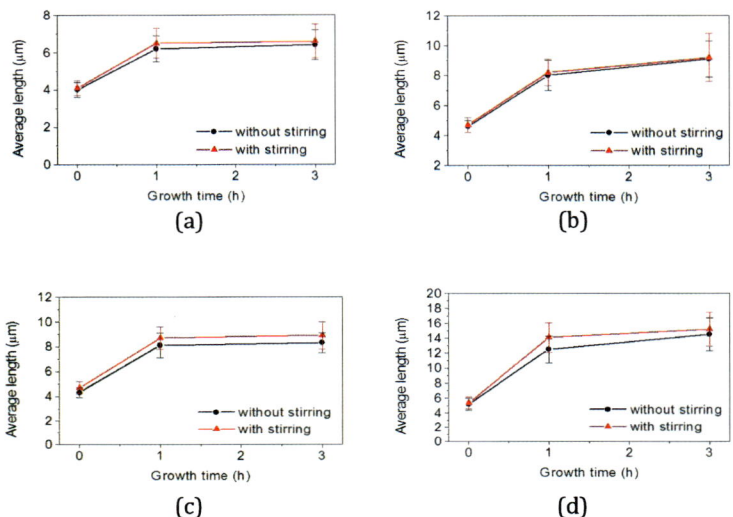

Figure 3.2 Measured average lengths of C_{60}NWs grown in the solutions agitated by magnetic bars with a length of 7 mm at (a) 5°C, (b) 10°C, (c) 15°C, and (d) 20°C. (Reprinted with permission from Hotta,[3] © 2008, World Scientific Publishing Company.)

The measured length growth rates are plotted as a function of $1/T$, as shown in the Arrhenius plot Fig. 3.3. The length growth data for the initial growth time of 5 min was used. Assuming the whisker length growth rate is proportional to $\exp(-\Delta E/RT)$, the activation energy of the whisker growth is calculated to be $\Delta E = 52.8$ kJ/mol from the slope of the fitted curve, where R is the gas constant.[4] This activation energy, 52.8 kJ/mol, is approximately four times greater than that of 13.1 kJ/mol for the diffusion of C_{60} in a mixed solvent of toluene and acetonitrile (4:1 v/v), which was obtained by using a microelectrode voltammeter.[5] This result supports the conjecture that the chemical reaction on the whisker surface governs the growth of C_{60}NWs. A high energy required to decompose the C_{60} molecules bonded with the solvent molecules must be necessary for the whisker growth in the desolvation process. Hence, the above high activation energy is supposed to include a high desolvation energy.

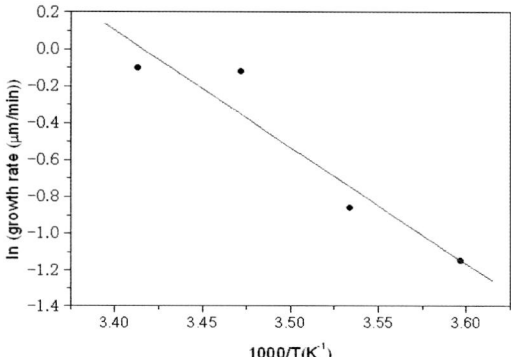

Figure 3.3 Temperature dependence of the length growth rate of C_{60}NWs. (Reprinted with permission from Hotta,[3] © 2008, World Scientific Publishing Company.)

3.4 EFFECT OF THE SOLVENT RATIO ON THE GROWTH OF C_{60}NWs

The growth of C_{60}NWs was examined for various volume mixing ratios of the C_{60}-saturated toluene solution and IPA, and the volume mixing ratios (=solvent ratios) were set to 1:9, 1:7, 1:5, 1:3, 1:1, 3:1, 5:1, 7:1, and 9:1. In this experiment, the total volume of C_{60}-saturated toluene solution and IPA was constant and fixed to 9 mL while only the solvent ratio was changed. The length and the diameter of C_{60}NWs were measured for the growth time of 24 h at 20°C. The length growth of C_{60}NWs changed depending on the above solvent ratios (Fig. 3.4). The longest C_{60}NWs with an average length of 7.7 ± 1.9 μm and an average diameter of 483 ± 95 nm were synthesized for the solvent ratio of 1:1. When the solvent ratio was 1:3, very short C_{60}NWs and a few long C_{60}NWs appeared, and their average length was 2.5 ± 1.0 μm and average diameter was 402 ± 107 nm. C_{60}NWs became shorter and small granular particles increased with increasing concentration of IPA. For the solutions with the solvent volume ratios greater than 3, C_{60}NWs were not formed and a few bulky and granular C_{60} crystals appeared. The formation of C_{60} precipitates became more difficult with decreasing concentration of IPA.

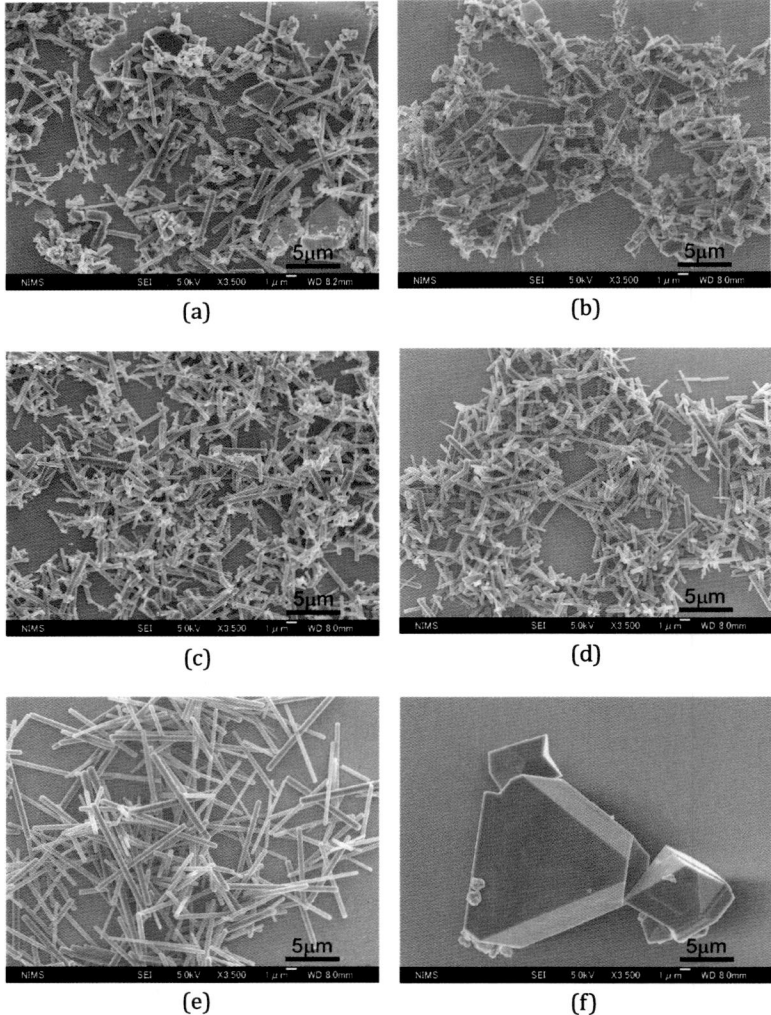

Figure 3.4 Scanning electron microscopy images of the C_{60} precipitates obtained for various volume mixing ratios of the C_{60}-saturated toluene solution and IPA. The volume mixing ratio of the C_{60}-saturated toluene solution and IPA is (a) 1:9, (b) 1:7, (c) 1.5, (d) 1:3, (e) 1:1, and (f) 3:1.[6]

Ringor et al. reported that thin and uniform C_{60} nanotubes, which are the tubular C_{60}NWs, can be prepared using C_{60}-saturated pyridine solutions and IPA with their volume mixing ratios of 1:10 and 1:9.[7] This result shows that the optimum mixing ratio of the C_{60}-saturated solution and IPA changes depending on the solvents used in the synthesis to obtain good-quality crystalline C_{60} nanofibers. As shown above, the mixing ratio of C_{60}-saturated toluene solution and IPA is found to influence the growth of C_{60}NWs.

3.5 EFFECT OF WATER ON THE GROWTH OF C_{60}NWs

The effect of a small amount of water contained in IPA on the growth of C_{60}NWs was investigated by the use of IPA-added pure distilled water. The mixing ratio of C_{60}-saturated toluene solution and IPA was 1:1 in volume. The water content in the as-received IPA was measured to be 0.017 mass %, and IPA solutions containing water of 0.38, 0.65, 0.89, 1.3, 2.5, 3.8, 5.0, 6.1, 7.1, 8.2, 9.4, 10.7, and 11.5 mass % were prepared for the C_{60}NW growth experiment. The water content in the as-received IPA and the above IPA solutions of water was measured by the use of a Karl Fischer titrator.

The length of C_{60}NWs was observed to change depending on the water content in IPA. C_{60}NWs could be formed when IPA containing 0.017–1.3 mass % water was used. When IPA containing 2.5 mass % water was used, C_{60}NWs with lengths greater than 10 μm could be synthesized in 1 day. But after 7 days from the start of the synthesis, C_{60}NWs were found to be partly decomposed and many bulky and granular C_{60} crystals appeared (Fig. 3.5). When IPA containing more than 3.8 mass % water was used, no C_{60}NWs were formed, although granular C_{60} crystals precipitated.

The length of C_{60}NWs increased with increasing water content in IPA up to 2.5 mass % water (Fig. 3.6). The chemical properties and the cluster structure of solvent molecules must be changed by the addition of water. A strong dipole–dipole interaction among solvent molecules and water molecules is conjectured to influence the growth and morphology of C_{60}NWs. It is also inferred that water can act as a catalyst to lower the desolvation energy of C_{60} molecules combined with solvent molecules.

(a)   (b)

Figure 3.5 Optical micrographs of C_{60}NWs prepared by using IPA containing 2.5% water. Growth period of (a) 1 day and (b) 7 days.[6]

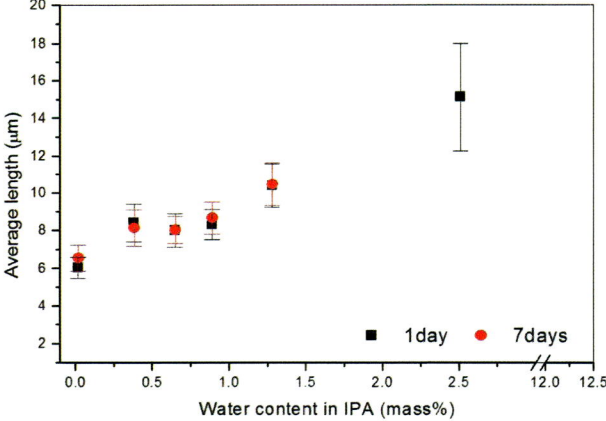

Figure 3.6 Average lengths of C_{60}NWs measured as a function of water content in IPA for the growth periods of 1 day and 7 days.[6]

The short C_{60}NWs shown in Fig. 3.5 and also in Fig. 3.7 will be applied to the electron collector electrodes of solar cells.[8]

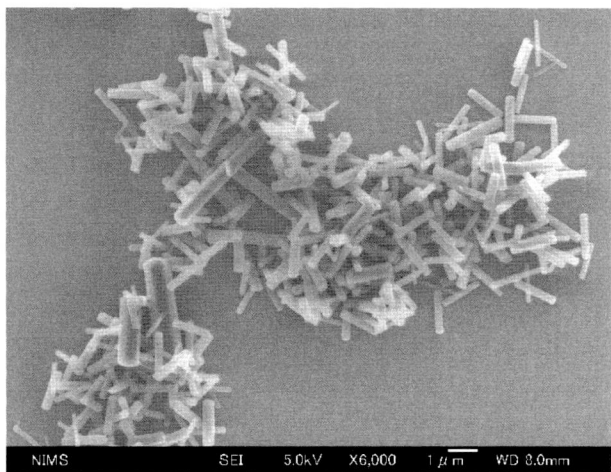

Figure 3.7 Short C_{60}NWs prepared by using the growth control method.

3.6 CONCLUSIONS

(1) The growth rate of C_{60}NWs increases with increasing solution temperature between 5 and 20°C. C_{60}NWs grow fastest in their initial growth stage, with a high degree of supersaturation.

(2) The growth rate of C_{60}NWs is scarcely influenced by the convection of solution, suggesting that the chemical reaction on the whisker surface governs the whisker growth.

(3) The activation energy of 52.8 kJ/mol was measured in the growth process of C_{60}NWs. This energy is supposed to include a high desolvation energy of the C_{60} molecules combined with the solvent molecules.

(4) C_{60}NWs were formed for the volume mixing ratios of the C_{60}- saturated toluene solution and IPA between 1:9 and 1:1. An optimum volume mixing ratio of toluene and IPA exists for the synthesis of good-quality C_{60}NWs.

(5) The length of C_{60}NWs increased with increasing water content in IPA in the range of 0.017–1.3 mass %. But when IPA with

the water content higher than 2.5 mass % was used, bulky and granular C_{60} crystals appeared instead of C_{60}NWs.

References

1. K. Miyazawa, A. Obayashi, and M. Kuwabara, *J. Am. Ceram. Soc.*, **84**, 3307 (2001).
2. K. Miyazawa, Y. Kuwasaki, A. Obayashi, and M. Kuwabara, *J. Mater. Res.*, **17**, 83 (2002).
3. K. Hotta and K. Miyazawa, *NANO*, **3**, 355 (2008).
4. W. Seifert, M. Borgstom, K. Deppert, K. A. Dick, J. Johansson, M. W. Larsson, T. Martensson, N. Skold, C. P. T. Svensson, B. A. Wacaser, L. R. Wallenberg, and L. Samuelson, *J. Cryst. Growth*, **272**, 211 (2004).
5. M. Wei, H. Luo, N. Li, S. Zhang, and L. Gan, *Microchem. J.*, **72**, 115 (2002).
6. K. Hotta and K. Miyazawa, *J. Phys.: Conf. Ser.*, **159**, 012021 (2009).
7. C. L. Ringor and K. Miyazawa, *Diam. Relat. Mater.*, **17**, 529 (2008).
8. P. R. Somani, S. P. Somani, and M. Umeno, *Appl. Phys. Lett.*, **91**, 173503 (2007).

Chapter 4

PREPARATION AND CHARACTERIZATION OF FULLERENE DERIVATIVES AND THEIR NANOWHISKERS

Shigeo Nakamura,[a] Kun'ichi Miyazawa,[b] and Tadahiko Mashino[a]

[a] *Department of Pharmaceutical Sciences, Faculty of Pharmacy, Keio University, 1-5-30, Shiba-koen, Minato-ku, Tokyo 105-8512, Japan*

[b] *National Institute for Materials Science, 1-1, Namiki, Tsukuba, Ibaraki 305-0044, Japan*

mashino-td@pha.keio.ac.jp; miyazawa.kunichi@nims.go.jp

Syntheses of three types of typical fullerene derivatives — malonic acid-type, pyrrolidine-type, and metal complex — are described. Nanowhiskers containing these fullerene derivatives were prepared from a mixture of C_{60} and fullerene derivatives. These nanowhiskers were single crystalline, indicating that the fullerene derivatives formed a solid solution with C_{60}. However, incorporation of fullerene derivatives into the C_{60} matrix in fullerene nanowhiskers was inhibited by large functional groups of fullerene derivatives. The use of fullerene derivatives for nanowhiskers is a potentially convenient method for providing a novel function to fullerene nanowhiskers.

Fullerene Nanowhiskers
Edited by Kun'ichi Miyazawa
Copyright © 2012 Pan Stanford Publishing Pte. Ltd.
www.panstanford.com

4.1 INTRODUCTION

Fullerene nanowhiskers (FNWs) are comparable to carbon nanotubes (CNTs) because they are both low-dimensional carbon nanomaterials. CNTs are potential novel materials for fuel cell electrodes, electron microscope probes, field emission displays, and other applications. Highly purified CNTs are necessary for these purposes; however, purification is difficult because CNTs are insoluble.

Incorporation of functional groups into CNTs is helpful in controlling the properties of CNTs or in adding new functions. However, derivatization of CNTs is also difficult because the reactivities of CNTs are lower than those of fullerenes. Even in the cases of successful derivatization, the incorporation ratios and the substitution positions are difficult to analyze.

Fullerenes can easily be incorporated into functional groups, and the identification of their derivatization is more established than that of CNTs. For the incorporation of functional groups, FNWs have an advantage over CNTs. In this chapter, the preparation of fullerene derivatives and FNWs containing fullerene derivatives is described.

4.2 SYNTHESIS OF FULLERENE DERIVATIVES

Many types of fullerene derivatives have been synthesized by hydrogenation, radical addition, nucleophilic addition, cycloaddition, and other methods. Hydroxylated fullerenes (fullerenols) are well-known fullerene derivatives that can be prepared by heating C_{60} in toluene in the presence of potassium hydroxide.[1] The product is not a homogeneous compound, however, because fullerenols are mixtures of multiple adducts. Furthermore, the characteristics of fullerenes cannot be utilized, since the wide resonance system of fullerenes is broken into multiple adducts. Monosubstituted fullerene derivatives, therefore, are more suitable for the use of FNWs containing fullerene derivatives. In this section, the syntheses of three types of C_{60} monoadducts are described.

4.2.1 Malonic Acid Derivatives

Adducts of C_{60} with diethyl malonate are easily obtained from the reaction of C_{60} and diethyl bromomalonate, with sodium hydride as a base in toluene (Fig. 4.1).[2] This cyclopropanation is a clean and mild reaction, and the reaction mixture contains unreacted C_{60}, monoadducts, bisadducts, and higher adducts. To obtain a monoadduct (**1**) in a good yield, an equivalent of diethyl bromomalonate is necessary. The monoadduct can easily be separated by silica gel column chromatography.

Ethyl ester (**1**) is converted to carboxylic acid by sodium hydride in toluene at 60°C, followed by methanol treatment.[3] Malonic acid derivatives are widely used as water-soluble fullerene derivatives.

Cyclopropanation proceeds similarly in the case of the reaction of C_{60} with α-halo esters or α-halo ketones such as methyl 2-acetyl-2-chloroacetate, 2-chloro-2-phenylacetophenone, and 2-bromoacetophenone.[2]

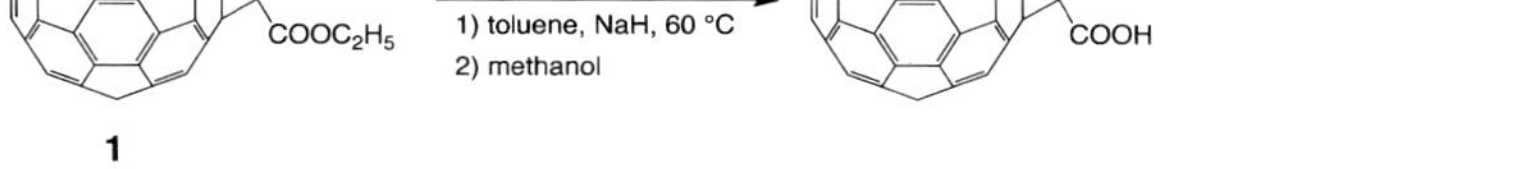

Figure 4.1 Synthesis of malonic acid derivatives of C_{60}.

4.2.2 Pyrrolidine Derivatives

The [6,6] double bonds of C_{60} behave as dienophiles in pericyclic reactions. Diels–Alder reactions can proceed from C_{60} and dienes. For example, C_{60} and cyclopentadiene react to give adducts in high yield at room temperature.[4]

Azomethine ylides (Fig. 4.2) are typical 1,3-dipoles that can react as a four-electron component to form five-membered rings in a pericyclic reaction. Pyrrolidine derivatives are synthesized from the reaction of C_{60} and an azomethine ylide generated from sarcosine (*N*-methylglycine) and aldehyde in good yield (Fig. 4.3).[5] This reaction is one of the most-used methods of fullerene derivatization. The *N*-methylpyrrolidine derivative **2** is produced when formaldehyde is used as an aldehyde, and **3** and **4** are also synthesized from the corresponding aldehydes.

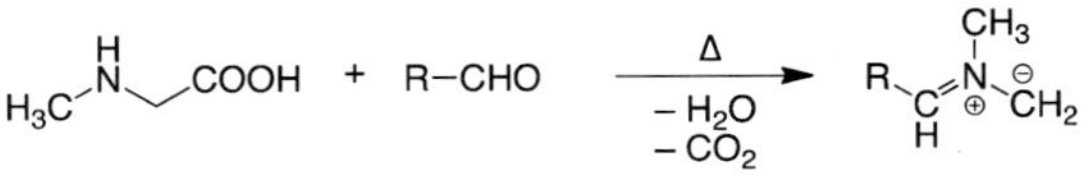

Figure 4.2 Production of azomethine ylides.

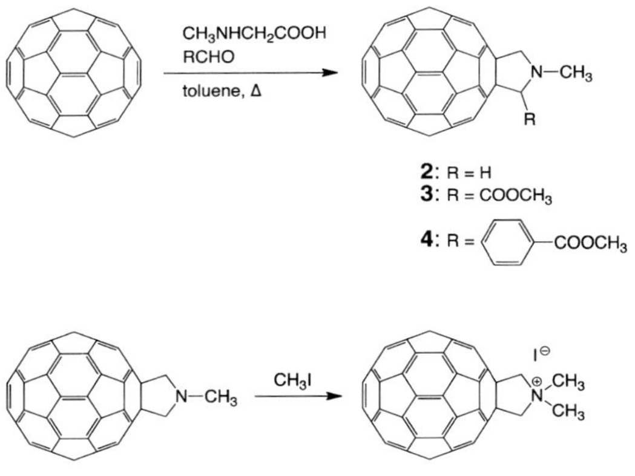

2: R = H
3: R = COOCH$_3$
4: R = ⟨phenyl⟩–COOCH$_3$

Figure 4.3 Synthesis of pyrrolidine derivatives of C_{60}.

Although the basicity and nucleophilicity of **2** are lower than those of *N*-methylpyrrolidine,[6] a quaternary ammonium derivative is produced from **2** and methyl iodide (Fig. 4.3).

4.2.3 Metal Complexes

Fullerenes have a characteristic electron-deficient olefin; therefore, double bonds of C_{60} coordinate to metals to form metal complexes. The binding mode of C_{60} to the transition metal is η^2-fashion in that two carbon atoms coordinate to the metal. A [6,6] double bond binds to the metal because of the electron deficiency of [6,6] double bonds; indeed, X-ray structure analysis showed the η^2-binding of the metal to C_{60}.

$((C_6H_5)_3P)_2Pt(\eta^2\text{-}C_{60})$ (**5**) is synthesized from the reaction of equimolar amounts of $((C_6H_5)_3P)_2Pt(\eta^2\text{-}C_2H_4)$ with C_{60} (Fig. 4.4).[7] Similar reactions also proceed with $((C_2H_5)_3P)_4Ni$ or $((C_2H_5)_3P)_2Pd(\eta^2\text{-}CH_2=CHCOOCH_3)$.

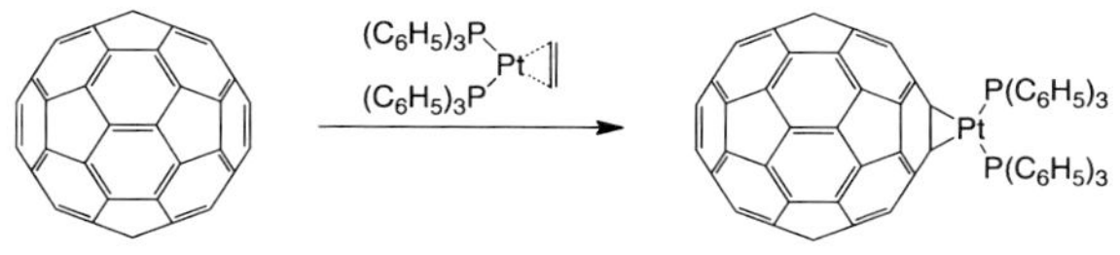

Figure 4.4 Synthesis of platinum complex of C_{60}.

4.3 FNWs CONTAINING FULLERENE DERIVATIVES

4.3.1 Preparation of FNWs Containing Fullerene Derivatives

C_{60} is relatively soluble in aromatic hydrocarbons. The solubility of C_{60} is about 3 mg/mL in toluene and about 5 mg/mL in *m*-xylene. The solubility of fullerene derivatives **14** is similar to that of C_{60}, and they are soluble in *m*-xylene up to 4–7 mg/mL.

FNWs containing **1–4** were prepared by using the liquid–liquid interfacial precipitation (LLIP) method described in previous chapters, using *m*-xylene as a good solvent and isopropyl alcohol as a poor solvent.[8] The ratios of **1– 4** to C_{60} were varied from 5 to 33 mol % in the starting composition. The contents of fullerene derivatives in FNWs were determined by high-performance liquid chromatography analysis. The incorporation efficiency of fullerene derivatives into FNWs was decreased when there were larger amounts of derivatives in the starting composition (Fig. 4.5). In the case of **4**, no crystals were obtained in the condition of 33 mol % fullerene derivative, suggesting that the dissolution of fullerene derivative molecules into the matrix of C_{60}NWs became more difficult as the size of substituent was increased.

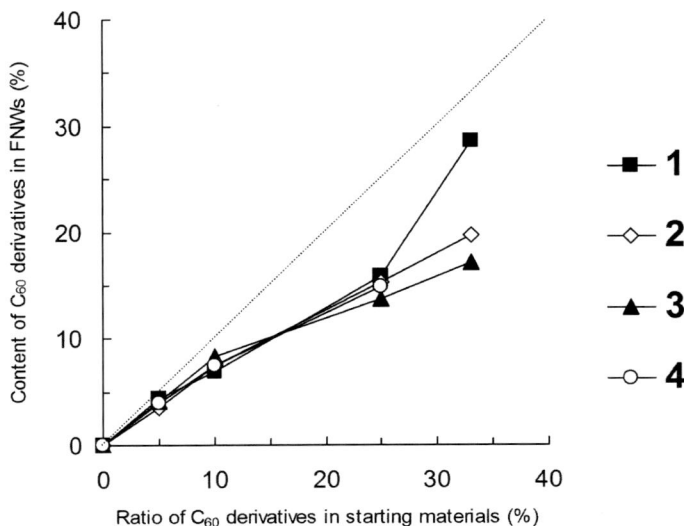

Figure 4.5 Composition of FNWs containing 1–4. (Reprinted with permission from Miyazawa et al.,[8] © 2009, IOP Publishing Ltd.)

4.3.2 Properties of FNWs Containing Fullerene Derivatives

A diversity of fullerene crystal morphologies was observed in FNWs containing fullerene derivatives depending on the starting composition (Figs. 4.6 and 4.7). Thicker crystals were obtained when the ratio of fullerene derivatives was higher or the size of the substituent was larger, as in the case of composition analysis. FNWs containing fullerene derivative **1** were observed by transmission electron microscopy (TEM) (Fig. 4.8).[9] The spots of a selected-area electron diffraction pattern (SAEDP) indicated that FNWs were single crystalline and were indexed by a hexagonal lattice from the fast Fourier transform pattern for a high-resolution TEM image. This result suggested that **1** formed a solid solution with C_{60} in FNWs. The hexagonal structure of FNWs was stabilized by **1** incorporated into the matrix of C_{60} nanowhiskers.

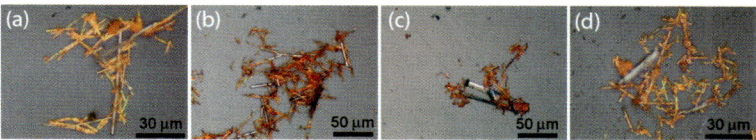

Figure 4.6 Optical micrographs of FNWs containing 1-4: (a) 1 = 4.3 mol %, (b) 2 = 3.5 mol %, (c) 3 = 4.1 mol %, and (d) 4 = 4.0 mol %. (Reprinted with permission from Miyazawa et al.,[8] © 2009, IOP Publishing Ltd.)

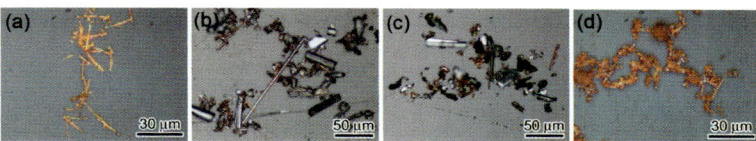

Figure 4.7 Optical micrographs of FNWs containing 1-4: (a) 1 = 29 mol %, (b) 2 = 20 mol %, (c) 3 = 17 mol %, and (d) 4 = 15 mol %. (Reprinted with permission from Miyazawa et al.,[8] © 2009, IOP Publishing Ltd.)

FNWs containing 7.8 mol % of the bisadduct of **3** were also prepared by using the LLIP method.[9] These FNWs showed a hexagonal unit cell, indicating the formation of an ordered solid solution of fullerenes. The use of the bisadduct led to the efficient incorporation of functional groups into FNWs.

The malonate derivative **1** was incorporated into FNWs with an excellent yield (Fig. 4.5); therefore, FNWs composed of 100% of **1** were successfully prepared by using the LLIP method.[10] The center-to-center distance (D) between two adjoining C_{60} cages of **1** was close to that of the C_{60}NWs, indicating that the C_{60} cages in nanowhiskers of **1** are linked along the whisker growth axis, with no malonate functional groups between the adjoining C_{60} cages (Fig. 4.9).

FNWs containing 1 mol % of the platinum complex **5** were successfully prepared by using the LLIP method.[11] Long FNWs, however, were not obtained in the case of 3 mol % of **5**, which indicated that the growth of FNWs was suppressed because the substituent of **5** containing a total of six benzene rings was large, as seen in molecular models (Fig. 4.10). A substituent that is too large tends to inhibit the formation of FNWs.

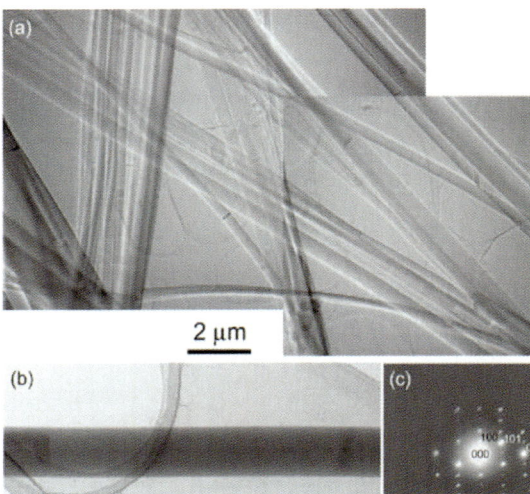

Figure 4.8 (a, b) TEM images for FNWs containing 4.2 mol % of 1. (c) SAEDP indexed by the hexagonal system for nanowhiskers. (Reprinted with permission from Miyazawa et al.,[9] © 2006, Institute of Nuclear Chemistry and Technology.)

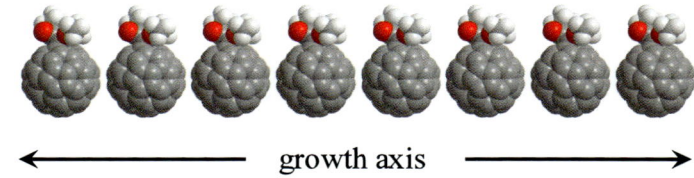

Figure 4.9 Schematic diagram of the nanowhisker 1.

Incorporating fullerene derivatives into FNWs is a convenient method of changing the properties of FNWs. The difficulty is that the efficiency of the incorporation decreases when the ratio of fullerene derivatives is higher or the substituent is larger. The incorporation of fullerene derivatives with highly polar groups, such as malonic acid derivatives and pyrrolidinium quaternary ammonium derivatives, is also expected. FNWs containing fullerene derivatives may produce novel materials related to carbon nanomaterials.

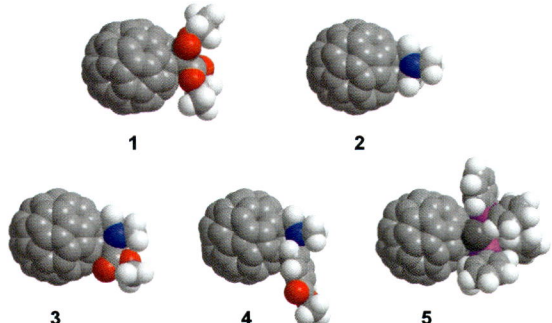

Figure 4.10 Comparison of molecular sizes of fullerene derivatives 1–5 by space-filling models.

4.4 CONCLUSION

A number of methods used for the synthesis of fullerene derivatives have been established. The preparation of FNWs containing malonate-type and pyrrolidine-type fullerene derivatives was successfully achieved by using the LLIP method. Fullerene-derivative-incorporated FNWs are promising nanomaterials for various applications.

References

1. A. Naim and P. B. Shevlin, *Tetrahedron Lett.*, **33**, 7097 (1992).
2. C. Bingel, *Chem. Ber.*, **126**, 1957 (1993).
3. I. Lamparth and A. Hirsch, *J. Chem. Soc., Chem. Commun.*, 1727 (1994).
4. V. M. Rotello, J. B. Howard, T. Yadav, M. M. Conn, E. Viani, L. M. Giovane, and A. L. Lafleur, *Tetrahedron Lett.*, **34**, 1561 (1993).
5. M. Maggini, G. Scorrano, and M. Prato, *J. Am. Chem. Soc.*, **115**, 9798 (1993).
6. F. D'Souza, M. E. Zandler, G. R. Deviprasad, and W. Kutner, *J. Phys. Chem. A*, **104**, 6887 (2000).
7. P. J. Fagan, J. C. Calabrese, and B. Malone, *Science*, **252**, 1160 (1991).

8. K. Miyazawa, R. Kato, K. Saito, T. Kizuka, T. Mashino, and S. Nakamura, *J. Phys.: Conf. Ser.*, **159**, 012007 (2009).
9. K. Miyazawa, J. Minato, T. Mashino, S. Nakamura, M. Fujino, and T. Suga, *NUKOEONIKA*, **51**, S41 (2006).
10. K. Miyazawa, T. Mashino, and T. Suga, *J. Mater. Res.*, **18**, 2730 (2003).
11. K. Miyazawa and T. Suga, *J. Mater. Res.*, **19**, 2410 (2004).

Chapter 5

VERTICALLY ALIGNED C_{60} MICROTUBE ARRAY

Cha Seung Il

Energy-Conversion Device Research Center, Korea Electrotechnology Research Institute,
70 Boolmosangil, Changwon 641-120, Korea
sicha@keri.re.kr

The liquid–liquid interfacial precipitation (LLIP) process can be modified as kinetic control of precipitation of C_{60} crystals is possible. In the modified LLIP process, fast precipitation condition gives fine nanowhiskers while slow precipitation induces microtubes. When the precipitation rate becomes very low, C_{60} crystals nucleate heterogeneously on the separating membrane and grow vertically with a well-developed hexagonal-shaped cross section and <110> crystallographic direction.

5.1 INTRODUCTION

C_{60} nanowhiskers, prepared by using the liquid–liquid interfacial precipitation (LLIP) process, have opened a new era of the application of fullerene-based materials including pure fullerenes and chemically modified fullerenes with its ability to fabricate

Fullerene Nanowhiskers
Edited by Kun'ichi Miyazawa
Copyright © 2012 Pan Stanford Publishing Pte. Ltd.
www.panstanford.com

large quantity of high-quality crystals of C_{60} nanostructures.[1-12] At the same time, it encourages one to apply it to other organic-based materials for organic electronics or to biomaterials such as proteins and nucleic acids.[13-15] Strictly said, the LLIP process is a kind of recrystallization process used for purification of organic materials. However, it should be noted that the nucleation and growth mechanisms of the LLIP process, as well as other kinds of recrystallization processes, are not clear. These mechanisms may give us a strong tool for controlling morphologies of fullerenes or other numerous useful organic materials.[16,17]

The LLIP process for C_{60} fullerenes consists of simple steps: preparation of C_{60} solution; slow addition of 2-propanol or other kinds of alcohols, called precipitator, to form an interface between the C_{60} solution and the precipitator; and waiting for slow diffusion of the precipitator in the C_{60} solution to form nanowhiskers at certain temperatures. A typical LLIP process described above does not have kinetically controllable variables. However, the precipitation of C_{60} nanowhiskers from the C_{60} solution is a kinetic process governed by kinetic parameters such as supersaturation, temperature, and diffusion.

Table 5.1 Shape of C_{60} crystals according to solvent-precipitator-promotor combinations in kinetically controlled LLIP process.

Shape	Solvent for C_{60}	Promotor	Precipitation
Nanowhiskers	Toluene	2-Propanol	2-Propanol
	Toluene	2-Propanol	Ethanol
Micro-tubes	Toluene	2-Propanol	2-Propanol
	Toluene	—	2-Propanol
Nanoribbon	Toluene	Ethanol + distilled water (99/1 volume ratio)	Ethanol + distilled water (99/1 volume ratio)
Polyhedral particles	Toluene	Ethanol	Ethanol, 2-propanol
Thin plates	Benzene	2-Propanol	2-Propanol
Spherical precipitates	Pyridine	Distilled water	Distilled water

In this chapter, some new attempts for a kinetically controlled LLIP process, modified from a typical LLIP process, are introduced. Also, a vertically aligned C_{60} microtube array, a product of the kinetically controlled LLIP process, is shown. In addition, some questions, raised by the use of kinetically controlled LLIP process, related to nucleation and growth mechanisms are introduced.

5.2 KINETICALLY CONTROLLED LLIP PROCESS

As mentioned above, kinetically controllable variables are limited in typical LLIP process. However, by simple modification of a typical LLIP process, the mixing rate of the precipitator with the solution as well as the supersaturation of C_{60} fullerenes in the solution can be controlled — by changing the stacking sequence of the solution and the precipitator. Due to different densities, the C_{60} solution is placed as the bottom layer and the precipitator, generally alcohols, is placed as the upper layer during the LLIP process forming the interface between the C_{60} solution and the precipitator. [17]

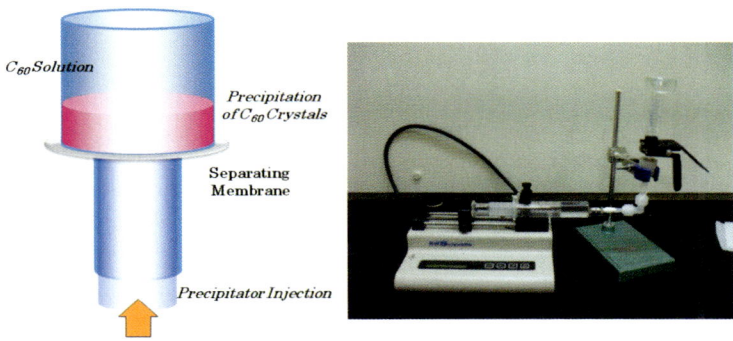

Figure 5.1 A schematic illustration (left) and picture of simple apparatus (right) for kinetically controlled LLIP process. In the apparatus, the precipitator is injected by a syringe pump and the separating membrane is anodized alumina of 200 nm pore size.

In order to change this sequence, the precipitators are injected from the bottom of the C_{60} solution in a controlled manner, as sketched in Fig. 5.1. The gravity and diffusion bring the injected precipitator into the whole C_{60} solution. If the injection rate of the precipitator is not faster than its mixing speed, then the mixing rate of the precipitator can be controlled by the injection rate of the precipitator. The supersaturation of C_{60} in the solution can be controlled by the addition of a small amount of alcohol, called promoter, that does not form precipitate. These modifications imply the separation of the C_{60} solution and the precipitator before the process and then mixing them in a controlled manner.

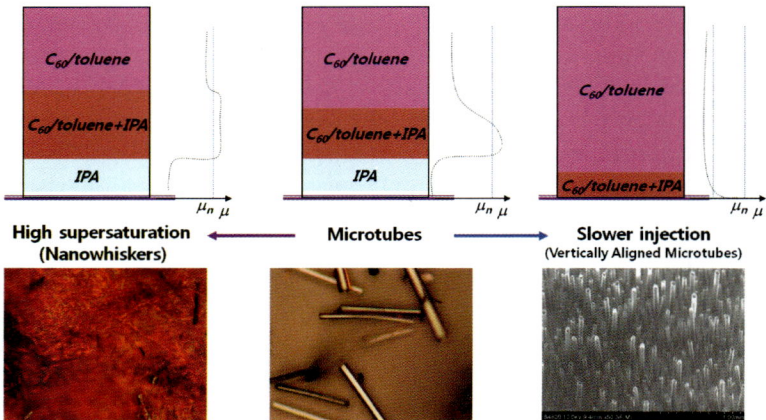

Figure 5.2 A schematic illustration for one-dimensional C_{60} crystals prepared by kinetically controlled LLIP process using toluene as solvent, 2-propanol as precipitator and promoter (up) and their morphologies (down) including nanowhiskers (left), suspended microtubes (middle), and vertically aligned microtube array (right).

The schematic illustration for C_{60} precipitation according to precipitator injection rate and supersaturation is given in Fig. 5.2 when the solvent of C_{60} solution is toluene and the precipitator and promoter are 2-propanol. In the figure, all precipitated C_{60} crystals have one-dimensional morphologies including nanowhiskers and microtubes. These phenomena can be easily explained by the classical nucleation theory: large amount of supersaturation induces rapid nucleation and slow growth, due to the consumption of C_{60} molecules for nucleation, and hence fine nanowhiskers, while lack of supersaturation induces large microtubes. One interesting fact is that when the injection rate is very high, such as sudden injection within several seconds, the nanowhiskers are formed without a promoter. As the injection rate decreases, the suspended microtubes change into a vertically aligned C_{60} microtube array due to heterogeneous nucleation of C_{60} on the separating membrane.

Another interesting feature in the precipitation of C_{60} crystals is that the promoter seems to determine the morphologies of the precipitated C_{60} crystals. In Fig. 5.3, when the promoter is ethanol and the precipitator is 2-propanol, C_{60} of polyhedral morphologies are precipitated while 2-propanol promoter and ethanol precipitator give nanowhiskers. Various combinations of promoters, precipitators, and solvents of the C_{60} solution induce several morphologies summarized in Table 5.1 and Fig. 5.4. It can be noted from this table that the resulting morphologies are not the same as those of a typical LLIP process. For instance, in a typical LLIP process, the C_{60}/pyridine solution gives fine nanowhiskers with 2-propanol precipitator when the C_{60}/pyridine solution is pretreated with ultraviolet or visible light.

 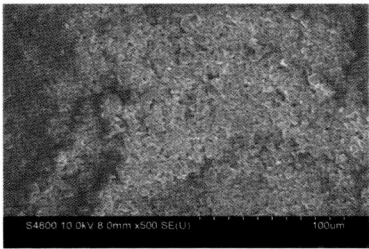

Figure 5.3 SEM micrographs of C_{60} crystals from the C_{60}/toluene solution with 2-propanol as the promoter and ethanol as the precipitator (left) as well as ethanol as the nucleator and 2-propanol as the precipitator (right). The precipitator injection rate was 0.1 mL/min.

5.3 VERTICALLY ALIGNED C_{60} MICROTUBES

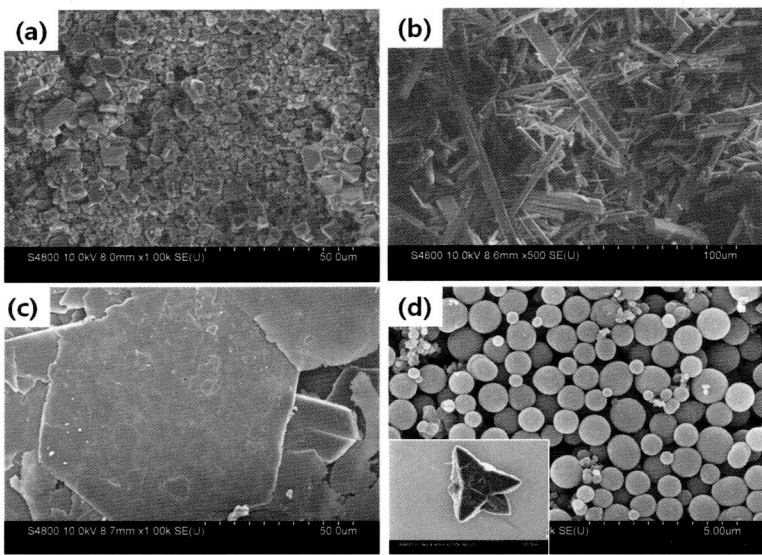

Figure 5.4 SEM micrographs of precipitated C_{60} crystals from the solution. (a) C_{60} polyhedral particles precipitated from the C_{60}/toluene solution using ethanol for supersaturation and precipitation, (b) C_{60} nanoribbons precipitated from the C_{60}/toluene solution using a mixture of ethanol and distilled water for supersaturation and precipitation, (c) C_{60} thin hexagonal plates precipitated from C_{60}/benzene solution using 2-propanol for supersaturation and precipitation, and (d) C_{60} spherical particles and star-shaped crystals precipitated from C_{60}/pyridine solution using water for supersaturation and precipitation.

Vertically aligned C_{60} microtubes are fabricated using toluene as the solvent and 2-propanol as the precipitator. When the injection rate of the precipitator is below 0.03 mL/min, the vertically aligned C_{60} microtubes are formed. Above that injection rate, the suspended C_{60} microtubes are formed within the mixed solution. Some micrographs of C_{60} microtubes are shown in Fig. 5.5, where hexagonal cross sections appear clearly. The density and diameter of C_{60} microtubes can be varied by changing the injection rate of the precipitator (Fig. 5.6).

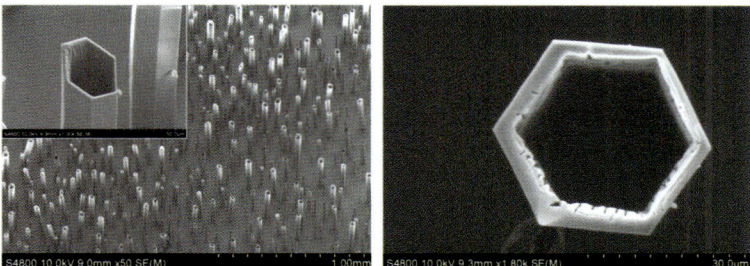

Figure 5.5 Side (left) and top views (right) of vertically grown C_{60} microtube arrays viewed with SEM. The inset in the left picture shows a close-up of C_{60} microtubes on the substrate.

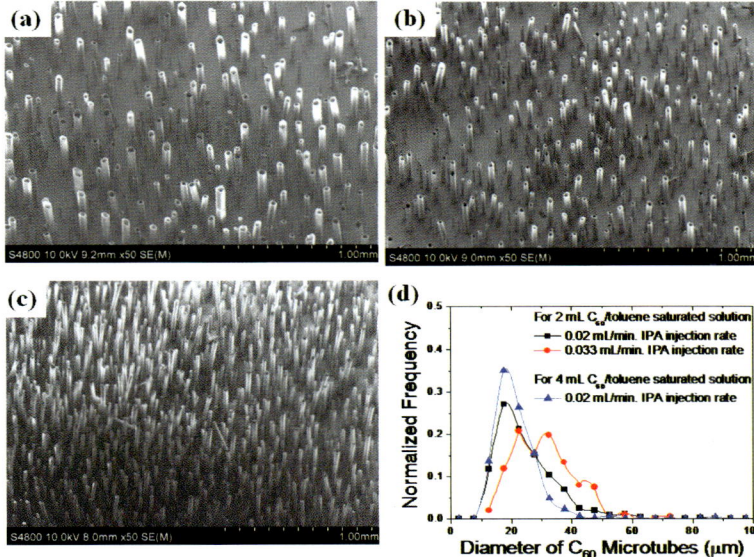

Figure 5.6 Side-view SEM micrographs of vertically aligned C_{60} microtube arrays fabricated by injecting 2-propanol at a rate of (a) 0.033 mL/min and (b) 0.02 mL/min into 2 mL of the C_{60}/toluene saturated solution, and (c) 0.02 mL/min into 4 mL of the C_{60}/toluene saturated solution. (d) Normalized distribution of the outer diameter of vertically aligned C_{60} microtubes with different injection rates of 2-propanol and amounts of the C_{60}/toluene solution. (Reprinted with permission Cha et al.,[17] © 2008, American Chemical Society.)

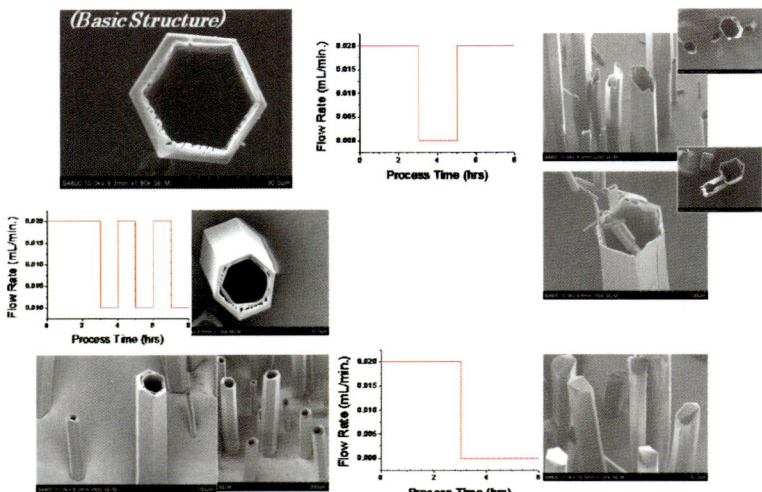

Figure 5.7 Changes of an inner structure of vertically aligned C$_{60}$ microtubes by modified injection of the precipitator. Basic structure (left, top) is changed into double-walled structure by stepwise injection (left down), into complex wall structure (right, up) by insertion of one incubation and reinjection, and into microrods (right, down) by injection and incubation.

One notable feature in the structure of vertically aligned C$_{60}$ microtubes is that the outer surface is smooth and the inner surface is serrated; in addition, it looks like something flows. The inner structure changes according to the injection profile: when 2-propanol is injected continuously until the vertically aligned C$_{60}$ microtubes are formed, which is actually 4 h of injection with a rate of 0.02 mL/min into 2 mL of saturated C$_{60}$ solution in toluene, and then holding 2-propanol injection and having a incubation time of ~4 h, the inner space of microtubes are filled and they turn into microrods. It gives an interesting fact that the outer surfaces are formed first and then the inner structures. Considering the serrated structures of the inner space in vertically aligned C$_{60}$ microtubes, the inner structures can be modified by changing the injection mode. When the 2-propanol injection mode is modified as the mode of reinjection after holding, one more wall is formed within the C$_{60}$ microtubes. When 2-propanol was injected in the stepwise mode, complex inner structures were formed, including small microtubes on the top of large microtubes and porous inner structures (Fig. 5.7). The changes in the inner structure of vertically aligned C_{60} microtubes due to modification of the injection mode give rise to a possibility that the inner structures are mudlike phases and the outer surfaces are rigid solids.

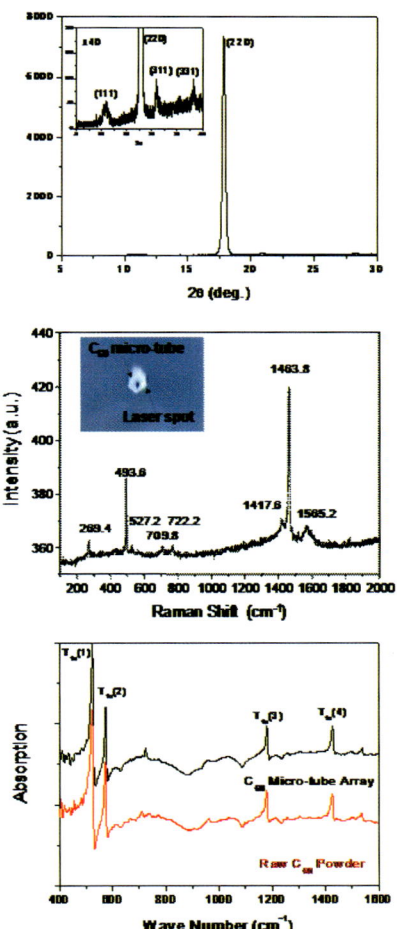

Figure 5.8 X-ray diffraction pattern (top), Fourier-transform infrared pattern (bottom), and Raman spectroscopy analysis (middle) of vertically aligned C_{60} microtubes. (Reprinted with permission from Cha et al.,[17] © 2008, American Chemical Society.)

The characterization of C_{60} microtubes indicates that the precipitated C_{60} microtubes are well crystallized without severe contamination or structural distortion (Fig. 5.8). In particular, X-ray diffraction pattern of C_{60} microtubes indicates that they are well crystallized as face-centered cubic crystal structures as sublimated C_{60} crystals and well aligned in <110> crystallographic direction. In addition, Raman spectroscopy and infrared spectroscopy of a vertically aligned C_{60} array and sublimated C_{60} crystals are almost identical, supporting the fact that there is little contamination of the solvent or the precipitator.

Above results including growth condition and structural characteristics can be interpreted by classical nucleation and growth mechanisms mentioned earlier. However, it is not the case for C_{60} microtubes. According to the classical theory, the nuclei form on the separating membrane and continuously grow in perpendicular and longitudinal directions, with different growth rates as their crystallographic characteristics. Therefore, it is expected that the short process time after nucleation gives us very small C_{60} crystals regardless of their shapes on the separating membranes. However, actual observation is different: the vertically aligned C_{60} microtubes suddenly appear in almost their final size after some incubation time (Fig. 5.9).

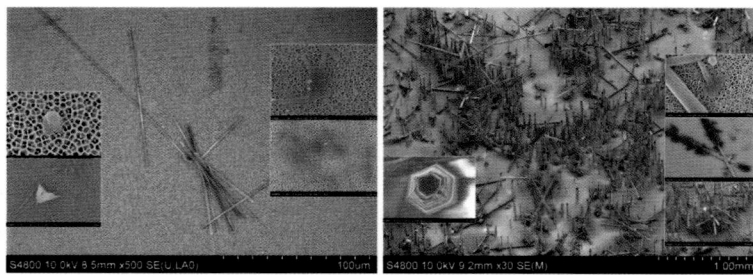

Figure 5.9 Micrographs of anodized alumina separating membrane after 1 mL of 2-propanol was injected into 3 mL of the C_{60}/toluene saturated solution with 0.02 mL/min injection rate (left) and after 2 mL of 2-propnaol was injected under the same conditions (right). Insets in both the figures show other places on the membranes.

5.4 CONCLUSION AND REMAINING PROBLEMS

In the classical nucleation theory, the nucleation rate is determined by the balance of the driving force for nucleation expressed as Gibbs free energy, i.e., thermodynamic energy difference between the solution and the precipitation, and surface energy expressed as the radius of curvature of the precipitation. In this theory, the binding energy between molecules forming the crystal nuclei determines the nucleation rate with kinetic parameters by controlling both the driving force and the surface energy. In many organic crystals, each molecule binds with weak interactions such as van der Waals forces or π–π stacking. The weak interaction between organic solute molecules implies a relatively strong interaction between the solvent and the solute.[18-22] Therefore, it is possible to form intermediate semisolid or high viscous phases composed of solvent, solute, and precipitator. Several examples are already shown in the case of some biomolecules such as proteins. Therefore, further studies should be proceeded to uncover the nucleation and growth mechanisms.

In summary, various shapes of C_{60} crystals can be prepared utilizing kinetically controlled LLIP process by changing combinations of solvents and precipitating alcohols. In particular, several well-crystallized one-dimensional C_{60} crystals, including nanowhiskers, microtubes, and vertically aligned microtubes, were precipitated from the solution by controlling the precipitation rate.

References

1. J. Minato and K. Miyazawa, *J. Mater. Res.*, **21**, 539, (2006).
2. K. Miyazawa, Y. Kuwasaki, A. Obayashi, and M. Kuwabara, *J. Mater. Res.*, **17**, 83 (2002).
3. K. Miyazawa, K. Hamamoto, S. Nagata, and T. Suga, *J. Mater. Res.*, **18**, 1096 (2003).
4. J. Minato and K. Miyazawa, *Carbon*, **43**, 2837 (2005).
5. M. Tachibana, K. Kobayashi, T. Uchida, K. Kojima, M. Tanimura, and K. Miyazawa, *Chem. Phys. Lett.*, **374**, 279 (2003).

6. M. Sathish, K. Miyazawa, and T. Sasaki, *Chem. Mater.*, **19**, 2398 (2007).
7. H. Liu, Y. Li, L. Jiang, H. Luo, S. Xiao, H. Fang, H. Li, D. Zhu, D. Yu, J. Xu, and B. Xiang, *J. Am. Chem. Soc.*, **124**, 13370 (2002).
8. L. Wang, B. Liu, S. Yu, M. Yao, D. Liu, Y. Hou, T. Cui, G. Zou, B. Sundqvist, H. You, D. Zhang, and D. Ma, *Chem. Mater.*, **18**, 4190 (2006).
9. Y. Jin, R. J. Curry, J. Sloan, R. A. Hatton, L. C. Chong, N. Blanchard, V. Stolojan, H. W. Kroto, and S. R. P. Silva, *J. Mater. Chem.*, **16**, 3715 (2006).
10. H.-X. Ji, J.-S. Hu, Q.-X. Tang, W.-G. Song, C.-R. Wang, W.-P. Hu, L.-J. Wan, and S.-T. Lee, *J. Phys. Chem. C*, **111**, 10498 (2007).
11. L. Wang, B. Liu, D. Liu, M. Yao, Y. Hou, S. Yu, T. Cui, D. Li, G. Zou, A. Iwasiewicz, and B. Sundqvist, *Adv. Mater.*, **18**, 1883 (2006).
12. K. Miyazawa, J. Minato, T. Mashino, T. Yoshii, T. Kizuka, R. Kato, M. Tachibana and T. Suga, in *Proceedings of the 2nd JSME/ASME International Conference on Materials and Processing 2005, Seattle, WA, June 19–22, 2005*, pp. (SMS23)-1–(SMS23)-4.
13. J. S. Heo, N. H. Park, J. H. Ryu, and K. D. Suh, *Adv. Mater.*, **17**, 822 (2005).
14. X. Zhang, X. Zhang, K. Zou, C.-S. Lee, and S.-T. Lee, *J. Am. Chem. Soc.*, **129**, 3527 (2007).
15. R. Taylor and D. R. M. Walton, *Chem. Soc. Rev.*, **23**(4), 243 (1994).
16. M. Sathish and K. Miyazawa, *J. Am. Chem. Soc.*, **129**, 13816 (2008).
17. S. I. Cha, K. Miyazawa, and J.-D. Kim, *Chem. Mater.*, **20**, 1667 (2008).
18. F. Cataldo, *Polym. Int.*, **48**, 143 (1999).
19. S. Toscani, H. Allouchi, J. Ll. Tamarit, D. O. Lopez, M. Barrio, V. Agafonov, A. Rassat, H. Szwarc, and R. Ceolin, *Chem. Phys. Lett.*, **330**, 491 (2000).
20. R. Ceolin, J. Ll. Tamarit, M. Barrio, D. O. Lopez, P. Espeau, H. Allouchi, and R. J. Papoular, *Carbon*, **43**, 417 (2005).
21. M. V. Korobov, E. B. Stukalin, A. L. Mirakyan, I. S. Neretin, Y. L. Slovokhotov, A. V. Dzyabchenko, A. I. Ancharov, and P. Tolochko, *Carbon*, **41**, 2743 (2003).
22. F. Michaud, M. Barrio, D. O. Lopez, J. Ll. Tamarit, V. Agafonov, S. Toscani, H. Szwarc, and R. Ceolin, *Chem. Mater.*, **12**, 3595 (2000).

Chapter 6

METAL-ION-INCORPORATED FULLERENE NANOWHISKERS AND SIZE-TUNABLE NANOSHEETS

Marappan Sathish and Kun'ichi Miyazawa

Fullerene Engineering Group, National Institute for Materials Science, 1-1, Namiki, Tsukuba, Ibaraki 305-0044, Japan
miyazawa.kunichi@nims.go.jp

This chapter deals with the liquid–liquid interfacial precipitation methods for the incorporation of metal ions such as Fe, Ni, and Ce into fullerene nanowhiskers. It was found that the incorporation of metal ions into fullerene nanostructures was affected by several parameters such as the nature of the solvent, concentration, and temperature. The X-ray diffraction pattern of the metal-ion-incorporated fullerene nanostructures reveals the face-centered cubic crystalline structure of the fullerene nanowhiskers. The Raman spectra of the metal-ion-incorporated fullerene nanostructures show light-induced polymerization of fullerene molecules in the nanostructures. In the end, this chapter deals with the selective

---

Fullerene Nanowhiskers
Edited by Kun'ichi Miyazawa
Copyright © 2012 Pan Stanford Publishing Pte. Ltd.
www.panstanford.com

precipitation of size tunable hexagonal fullerene nanosheets/disks. It is shown that the fullerene nanosheets/disks are a promising candidate for solar cell applications.

6.1 INTRODUCTION

Preparation methods and the effect of various parameters on the formation of fullerene nanostructures, particularly fullerene nanowhiskers and nanotubes, have been discussed in the previous chapters. As the readers would have noted, the liquid–liquid interfacial precipitation (LLIP) method has been identified as an effective method for the precipitation of fullerene nanostructures.[1] At this juncture it would be appropriate to introduce metal-ion-incorporated fullerene nanostructures prepared by using the LLIP method.

Carbon-based nanomaterials, particularly carbon nanotubues (single-walled and multiwalled) and metal or semiconductor nanoparticle-decorated carbon nanotubes, are well known for various practical applications.[2] Recently, attempts have been made to prepare various fullerene nanostructures doped or incorporated with various organic functional molecules, which enhance the use of fullerene nanostructure for various applications, owing to the integration of properties.[3,4] For the similar reason, the incorporation of inorganic nanostructures, such as metal, metal ions, metal oxides, and metal complexes, has also gained attraction due to their widespread applications. With this background, the idea of incorporating metal or metal ions into the fullerene nanostructures has been just conceived and few attempts have been initiated recently. In this chapter, more attention has been paid on the possible preparation methods, characterization, and effect of various parameters such as metal ions' concentration and temperature on the formation of metal-ion-incorporated fullerene nanowhiskers. In addition, preparation of various other interesting fullerene nanostructures, particularly fullerene nanosheets, will be discussed in detail.

6.2 RECENT FOCUS ON FULLERENE NANOMATERIALS

The recent thrust in the fullerene research is focused on the synthesis of various fullerene nanostructures such as nanowhiskers/nanotubes with active materials for application in various sectors.[5-7] Fullerene, fullerene derivatives, and fullerene-based nanostructures are good electron acceptors and can be used in solar cells as donor–acceptor systems.[8] In the case of fullerene nanowhiskers, the major drawback associated with them is the low electrical conductivity as compared to carbon nanotubes, which limits the application of fullerene nanowhiskers. However, the alkali-metal-doped C_{60} fullerides have been reported to exhibit superconductivity,[9] and the C_{60} nanowhiskers show a good electrical conductivity when compared to the pristine C_{60} powder. It is presumed that the electrical conductivity of the fullerene nanowhiskers or fullerene nanotubes can be improved by decreasing the diameter of the whiskers or high-temperature heat treatment. As expected, the studies on the conductivity of fullerene nanowhiskers[10-12] show that a whisker of C_{60} with a diameter of >10 and 8 μm has a resistivity of about 10^8–10^{10} and 10^5 Ω cm, respectively, whereas the pristine C_{60} crystals have the resistivity value in the range of 10^8–10^{14} Ω cm. Likewise, a fullerene whisker of 2.2 μm in diameter showed a resistivity of 0.037 Ω cm and a fullerene whisker of 2.9 μm in diameter showed a resistivity of 0.042 Ω cm after the heat treatment at 1100°C for 30 min in vacuum.

Furthermore, the conductivity of the C_{60} nanowhiskers could be increased by high-temperature heat treatment in vacuum. Again, it is speculated that the incorporation of transition metal ions such as Fe, Co, and Ni in the C_{60} nanowhiskers could also increase the electrical conductivity of the nanowhiskers to a greater extent and it can be used for various electrochemical applications such as anode support material for fuel cells and batteries. An attempt has been made to incorporate transition metal ions such as Fe, Ni, Ce, Ti, and V into C_{60} nanowhiskers and study their electrical conductivity.

6.3 PRECIPITATION OF FULLERENE NANOSTRUCTURES USING THE LLIP METHOD

As stated earlier, in the LLIP method, two types of solvents are involved. One of the solvents has a high solubility for fullerene, such as benzene, toluene, or *m*-xylene, and the second one has a low solubility for fullerene, such as isopropyl alcohol (IPA) or 1-butanol. In the case of preparation of metal-ion-incorporated fullerene nanostructures, first the metal ions are predissolved into the solvent having a low solubility for fullerene, such as IPA or 1-butanol, and then a saturated solution of fullerene was prepared in benzene, toluene, or *m*-xylene. Appropriate concentrations of $FeCl_3 \cdot 6H_2O$, $Ni(NO_3)_2 \cdot 6H_2O$, and $Ce(NO_3)_3 \cdot 6H_2O$ were used as the metal-ions' precursors for the incorporation of Fe, Ni, and Ce ions into the fullerene nanostructures,[13-15] respectively. The preparation method, characterization, and potential importance of the metal-ion-incorporated fullerene nanostructures will be discussed in the following sections separately.

6.3.1 Fe-Ion Incorporation in Fullerene Nanostructures

In the LLIP method, the conditions, i.e., the type of solvent and the concentration of metal ion precursor, vary according to the type of metal ion intended to be incorporated into the fullerene nanostructure. In the case of preparation of Fe-ion-incorporated fullerene, it should be kept in mind that the concentration of the Fe ions in the second solvent should be optimum. A high concentration of Fe ions in the solution results in no proper precipitation of fullerene nanostructures. Similarly, a low concentration of Fe ions in the solution results in no significant amount of incorporation. It was found that the amount of Fe-ion incorporation and the morphology of fullerene nanostructure differ with the solvent nature. Here a summary of the various solvent combinations and ratios used in the preparation of Fe-ion-incorporated fullerene nanostructures are listed. Table 6.1 gives the amount of Fe-ion-incorporated in the fullerene nanowhiskers, the nature of solvents, the initial concentration of Fe ions in the solutions, and the morphology of resulting fullerene nanostructures. It could be clearly seen that there is no direct correlation between the initial concentration of Fe ions and the amount of incorporated Fe ions in the nanostructures. The toluene solvent shows a higher amount of Fe-ion incorporation in the nanowhiskers compared to benzene and m-xylene. Among benzene and m-xylene solvents, benzene shows comparatively lower amount of Fe-ion incorporation. In the case of the benzene solvent, a high initial concentration of Fe ions in IPA (0.005 M) does not result in any precipitation of fullerene nanowhiskers, whereas

Table 6.1. Amount of Fe-ion incorporation into fullerene nanowhiskers at different experimental conditions (temperature = 5°C).

S. No.	Solvent I/ solvent II	Benzene/IPA	Toluene/IPA	m-Xylene/IPA
1	Initial concentration of Fe ions (moles)	0.001	0.005	0.005
2	Amount of Fe-ion incorporation (wt %)	0.3–05	>2.75	0.4
3	Morphology			

Fe ions in 1-butanol (0.005 M) results in the formation of fullerene nanowhiskers with 2.75 wt % Fe-ion incorporation. The observed results evidently indicate an interaction between Fe ions and fullerenes, which could be affected by the nature of the solvent molecules. However, the precise interaction and mechanism between the Fe ions and the fullerene molecules are indistinct from using the current characterization methods.

6.3.2 Ni-ION-INCORPORATED FULLERENE NANOWHISKERS

In this section, the incorporation of Ni ions into the fullerene nanowhiskers is discussed. The use of benzene–IPA in the LLIP method successfully produced nanoporous fullerene nanowhiskers with a high surface area.[16] The incorporation of active elements into these nanoporous materials will allow their dispersion in a higher amount with a better homogeneity. The dispersion of Ni ions on such nanoporous fullerene nanowhiskers is attractive in terms of its catalytic application. The preparation was very similar to the method described in the earlier section. For a typical preparation, 4.5 mL of 0.05 M $Ni(NO_3)_2.6H_2O$ containing the IPA solution was added into 1 mL of C_{60} saturated benzene solution. A detailed structural characterization of Ni-ion-incorporated fullerene nanowhiskers is as follows.

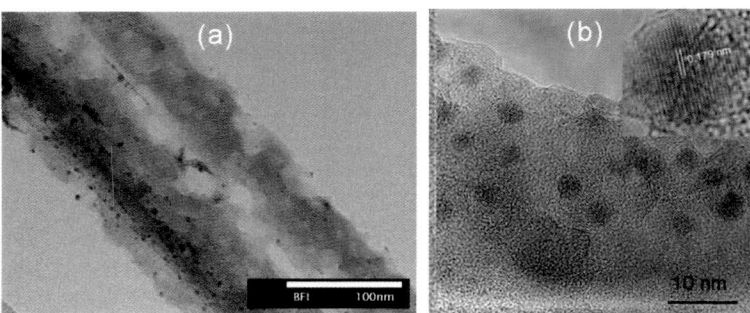

Figure 6.1 (a) TEM and (b) high-resolution TEM images of Ni-ion-incorporated fullerene. (Reprinted with permission from Sathish et al.,[14] © 2007, Elsevier.)

The microscopic images of the Ni-ion-incorporated C_{60} nanowhiskers[14] show both the tubular and nontubular structural morphologies such as metal-free nanoporous fullerene nanowhiskers.[16] Although the wall surfaces of the fullerene nanowhiskers incorporating Ni appear to be uniform at low magnifications in the observation by transmission electron microscopy (TEM), they show lopsided surfaces at high magnifications. However, the porous nature of the Ni-ion-incorporated C_{60} nanowhiskers does not resemble the metal-free nanoporous C_{60} nanowhiskers. The tubular nanoporous fullerene nanowhiskers have the inner tube diameter of ~100 nm,[16] whereas the Ni-ion-incorporated fullerene nanowhiskers show slightly smaller inner tube diameters.

Certainly, the presence of black particles in the tubular part of the nanowhiskers could be clearly seen from the TEM images (Fig. 6.1). The energy-dispersive X-ray spectroscopy (EDX) analysis confirms that the black spots are the incorporated Ni atoms in the nanowhiskers. The high-resolution TEM image (Fig. 6.1b) shows the lattice planes of the Ni particles dispersed over the C_{60} nanowhiskers. The observed lattice planes with the spacing of 0.179 nm (inset of Fig. 6.1b) for the black particles were attributed to Ni (200) planes.

6.3.3 Ce-Ion-Incorporated Fullerene Nanowhiskers

CeO_2-based materials are well known for oxygen storage by releasing and uptaking oxygen through redox processes by the Ce^{4+}/Ce^{3+} couple ($CeO_2 \leftrightarrow CeO_{2-x} + X/2\ O_2$). Due to this high oxygen storage capacity and high catalytic activity toward CO oxidation, they can be used as the electrocatalysts in direct methanol fuel cells (DMFCs). In the available commercial anode catalyst (Pt-Ru/C) used in the DMFCs, the role of Ru is to oxidize the CO molecules that are formed on the Pt surface during the methanol oxidation. CeO_2 used in the electrode materials will replace the Ru metal, which reduces the cost of the fuel cells. With this background, attempts have been made to prepare the C_{60} nanowhiskers containing Ce by using the LLIP method. The salient features of the Ce-ion-incorporated fullerene nanostructured materials are as follows: (1) C_{60} nanowhiskers themselves can act as the supports for the anode materials instead of commercial carbon materials; (2) a fine dispersion of Ce on the nanowhiskers will be achieved, which forms a high surface area for the fine dispersion of Pt to improve the cell performance.

An IPA solution containing $Ce(NO_3)_3 \cdot 6H_2O$ and a benzene solution saturated with C_{60} were used for the preparation of Ce-containing fullerene nanowhiskers.[15] Although the formation of porous Ce-ion-incorporated fullerene nanowhiskers was not observed in the as-prepared sample (Fig. 6.2a), the heating of the Ce-ion-incorporated nanowhiskers at 200°C for 2 h in vacuum results in the formation of porous nanowhiskers (Fig. 6.2b). However, high-temperature (>400°C) heat treatment destroys the whiskers' morphology. In order to locate the Ce ions in the fullerene nanowhiskers, the EDX mapping was carried out and the mapping image is shown in Fig. 6.2c. It can be clearly seen from the image that Ce ions just cover the surface of the fullerene nanowhiskers and the concentration is low in the inner part of the fullerene nanowhiskers.

Figure 6.2 SEM images of (a) as-prepared Ce-ion-incorporated fullerene nanowhiskers, (b) Ce-ion-incorporated fullerene nanowhiskers heat-treated at 250°C, and (c) scanning transmission electron microscopy mapping images of the Ce-ion-incorporated fullerene nanowhiskers. (Reprinted with permission from Sathish et al.,[15] © 2008, Springer.)

Even after increasing/decreasing the concentration of the Ce ions in the precursor solutions, the amount of the incorporated Ce ions in the fullerene nanowhiskers was low. It is speculated that the greater radius of the Ce ions compared to that of Fe and Ni ions limits the incorporation of ions into the fullerene nanowhiskers.

6.3.4 X-Ray Diffraction Pattern and Raman Spectra of the Metal-Ion-Incorporated Fullerene Nanostructures

The X-ray diffraction (XRD) patterns (Fig. 6.3a) of metal-ion (Fe, Ni, and Ce)-incorporated C_{60} nanowhiskers show the face-centered cubic (fcc) structure[17] with slightly broader lines compared to the pristine C_{60} powder owing to the particle size effect. There is a significant shift in the 2θ values (slightly higher 2θ values) for the Ce-ion-incorporated nanowhiskers compared to the other metal-ion-incorporated nanowhiskers. This shift can be attributed to the interaction between the Ce ions and the C_{60} molecules in the Ce-ion-incorporated nanowhiskers. In all the three cases, the peak broadening can be ascribed to the smaller particle size of the nanowhiskers compared to the bulk particle size of the pristine C_{60} powder. However, the XRD peaks for the Fe-ion-incorporated nanowhiskers are very broad and the peaks at a higher 2θ range are absent compared to the pristine C_{60} powder. This can be attributed to the low crystalline nature of the Fe-ion-incorporated nanowhiskers. In addition, there are no lines corresponding to Fe, Ni, or Ce due to the fine dispersion and low relative concentration of metal ions in the nanowhiskers. Therefore, the incorporation of metal ions in the C_{60} nanowhiskers does not affect the crystalline nature and the crystal structure. The calculated lattice constants for the Fe-, Ni-, and Ce-ion- incorporated fullerene nanowhiskers are in good agreement with the fcc crystalline structure.

The Raman spectra of the metal-ion-incorporated C_{60} nanowhiskers show a shift in the line at 1458 ± 1 cm^{-1} corresponding to the Ag(2) mode and few additional lines (Fig. 6.3b) compared to those of the C_{60} powder. This observation supports the possibility of polymerization of C_{60} molecules in the nanowhiskers.[18] However, the metal-ion (Fe, Ni, and Ce)-incorporated fullerene nanowhiskers show only fcc crystalline structure, contrary to the possibility of polymerization. Thus, it is speculated that a polymerization might have occurred during the Raman spectroscopic measurement due to the influence of the laser beam used for the measurement (532 nm). In the literature, photo-assisted polymerization of C_{60} molecules also shows a clear shift at 1459 cm^{-1}, which supports the above speculation.[19,20] It is worth noting here that the absence of (200) reflection in the XRD pattern for the nanowhiskers is the indication of free rotation of the C_{60} molecule in the nanowhiskers, which is not viable if the polymerization would have taken place beforehand.

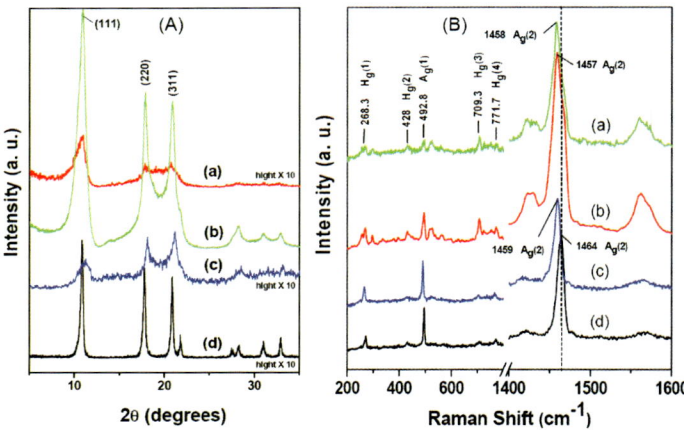

Figure 6.3 (a) XRD pattern and (b) Raman spectra of (a) Fe-, (b) Ni-, and (c) Ce-ion-incorporated fullerene nanowhiskers and (d) C_{60} powder.

6.3.5 Effect of Temperature

From our own experimental experiences in all these preparation of metal-ion-incorporated fullerene nanostructures, the ideal temperature for the optimal metal-ion-incorporated fullerene nanowhiskers was found to be 5°C among the studied temperatures of 5, 10, and 15°C. It is worth noting here that the most favorable temperature for the formation and growth of fullerene nanostructures, particularly fullerene nanowhiskers, was also 5°C. At high temperatures (>5°C), the solubility of fullerene will be higher, which might reduce the yield of precipitation.

6.4 FULLERENE NANOSHEETS

Among the fullerene nanostructures, fullerene nanosheets are one of the most interesting morphologies due to their very thin nature. To the best of our knowledge, only limited reports were available in the literature for the preparation of fullerene sheets and the reported fullerene sheets are very large in size. Shin *et al.* reported a vapor–solid process for the selective preparation of C_{60} disks and studied their photoconductivity.[21]

Figure 6.4 (a, b) Size tunable hexagonal fullerene nanosheets, (c) rhombus fullerene nanosheets, and (d) polygonal fullerene nanosheets prepared using different solvents in the LLIP method. (Reprinted with permission from Sathish et al., © 2007, American Chemical Society, and Sathish et al., © 2009, American Chemical Society.)

Similarly, Wang *et al.* have reported the preparation of fullerene nanosheets by using a simple solvent evaporation process and studied their pressure-induced amorphization behavior.[22] However, the thickness of the nanosheets prepared in these methods varies between a few hundreds in nanometers to a few micrometers.

Recently, the preparation of tunable size hexagonal nanosheets using the LLIP method was reported in the literature (Fig. 6.4).[23] Interestingly, by changing one of the solvents in the LLIP method, the size of the fullerene nanosheets was tuned from ~7 µm to 500 nm.

Also, the prepared sheets are uniform in shape (hexagonal) and size having very thin nature. Similarly, selective precipitation methods for the preparation of rhombus shape fullerene nanosheets and polygonal fullerene nanosheets were identified using different solvent systems in the LLIP method.[24]

It is surmised that by varying the solvent properties, one could selectively control the size and shape of the nanosheets. Besides, it would be interesting to incorporate various active materials into fullerene nanosheets and study their properties. Methodologies for selective precipitation of uniform size and flexible shape fullerene nanosheets with functional (organic and inorganic) molecules are emerging. It opens new rooms in materials science to study the fullerene-based solar cells and microelectronics for future applications.

6.5 SUMMARY

This chapter has focused on the preparation of metal-ion-incorporated fullerene nanostructures using the LLIP method. It is clear from the above observations that the solvents used for the precipitation play a crucial role both in the morphology of the fullerene nanostructures and in the amount of metal-ion incorporation. Thus, it is essential to select appropriate solvents to incorporate particular metal ions into the fullerene nanostructures. At this point it is difficult to derive a direct correlation between various parameters (metal ions initial concentration, morphology, and incorporated metal ions) and solvents, which needs more detailed experimental investigations. Fullerene nanosheets with uniform size and flexible shape are more attractive in recent days, and studies on the application of nanosheets are under progress.

References

1. K. Miyazawa, Y. Kuwasaki, A. Obayashi, and M. Kuwabara, *J. Mater. Res.*, **17**, 83 (2002).

2. V. Georgakilas, D. Gournis, V. Tzitzios, L. Pasquato, D. M. Guldi, and M. Prato, *J. Mater. Chem.*, **17**, 2679 (2007).

3. B. C. Thompson and J. M. J. Fréchet, *Angew. Chem. Int. Ed.*, **47**, 58 (2008).

4. H. Hoppe and N. S. Sariciftci, *J. Mater. Chem.*, **16**, 45 (2006).

5. K. Miyazawa, *J. Nanosci. Nanotechnol.*, **9**, 41 (2009).

6. M. Yao, B. M. Andersson, P. Stenmark, B. Sundqvist, B. Liu, and T. Wågberg, *Carbon*, **47**, 1181 (2009).

7. T. Hasobe, A. S. D. Sandanayaka, T. Wada, and Y. Araki, *Chem. Commun.*, 3372 (2008).

8. N. S. Sariciftci, L. Smilowitz, A. J. Heeger, and F. Wudl, *Science*, **258**, 1474 (1992).

9. A. F. Hebard, M. J. Rosseinsky, R. C. Haddon, D. W. Murphy, S. H. Glarum, T. T. M. Palstra, A. P. Ramirez, and A. R. Kortan, *Nature*, **350**, 600 (1991).

10. K. Miyazawa, J. Minato, H. Zhou, T. Taguchi, I. Honma, and T. Suga. *J. Eur. Ceram. Soc.*, **26**, 429 (2006).

11. M. P. Larsson, J. Kjelstrup-Hansen, and S. lucyszyn, *ECS Trans.,* **2**, 27 (2007).

12. K. Miyazawa, Y. Kuwasaki, K. Hamamoto, S. Nagata, A. Obayashi, and M. Kuwabara, *Surf. Interface Anal.*, **35**, 117 (2003).

13. M. Sathish and K. Miyazawa, *Nano*, **3**, 409 (2008).

14. M. Sathish, K. Miyazawa, and T. Sasaki, *Diam. Relat. Mater.*, **17**, 571 (2008).

15. M. Sathish, K. Miyazawa, and T. Sasaki, *J. Solid State Electrochem.*, **12**, 835 (2008).

16. M. Sathish, K. Miyazawa, and T. Sasaki, *Chem. Mater.*, **19**, 2398 (2007).

17. W. I. F. David, R. M. Ibberson, and T. Matsuo, *Proc. R. Soc. London A*, **442**, 129 (1993).

18. M. C. Martin, D. Koller, A. Rosenberg, C. Kendziora, and L. Mihaly, *Phys. Rev. B*, **51**, 3210 (1995).

19. M. Tachibana, K. Kobayashi, T. Uchida, K. Kojima, M. Tanimura, and K. Miyazawa, C*hem. Phys. Lett.*, **374**, 279 (2003).

20. A. M. Rao, P. C. Eklund, U. D. Venkatswaran, J. Tucker, M. A. Duncan, G. M. Bendele, M. Nuenez-Regueiro, I. O. Bashkin, E. G. Ponyatovsky, and A. P. Morovsky, *Appl. Phys. A*, **64**, 231 (1997).
21. H. S. Shin, S. M. Yoon, Q. Tang, B. Chon, T. Joo, and H. C. Choi, *Angew. Chem. Int. Ed.*, **47**, 693 (2008).
22. L. Wang, B. Liu, D. Liu, M. Yao, S. Yu, Y. Hou, B. Zou, T. Cui, G. Zou, B. Sundqvist, Z. Luo, H. Li, Y. Li, J. Liu, Z. Luo, H. Li, Y. Li, and J. Liu. *Appl. Phys. Lett.*, **91**, 103112 (2007).
23. M. Sathish and K. Miyazawa, *J. Am. Chem. Soc.*, **129**, 13816 (2007).
24. M. Sathish, K. Miyazawa, J. P. Hill, and K. Ariga, *J. Am. Chem. Soc.*, **131**, 6372 (2009).

Chapter 7

FABRICATION AND CHARACTERIZATION OF C_{60} FINE CRYSTALS AND THEIR HYBRIDIZATION

Akito Masuhara,[a] Zhenquan Tan,[b] Hitoshi Kasai,[c,d] Hachiro Nakanishi,[c] and Hidetoshi Oikawa[c]

[a] *Department of Organic Device Engineering, Yamagata University, 4-3-16 Jonan, Yonezawa 992-8510, Japan*
[b] *Joining and Welding Research Institute, Osaka University, 11-1 Mihogaoka, Ibaraki 567-0047, Japan*
[c] *Institute of Multidisciplinary Research for Advanced Materials, Tohoku University, 2-1-1 Katahira, Aoba-ku, Sendai 980-8577, Japan*
[d] *Precursory Research for Embryonic Science and Technology, Japan Science and Technology, 4-1-8, Honcho, Kawaguchi, Saitama 332-0012, Japan*

masuhara@tagen.tohoku.ac.jp

The preparation of fullerene fine crystals with a uniform size and shape would lead to a new class of materials with specifically controlled electronic state. To this end, we have successfully fabricated shape and size-controlled C_{60} fine crystals using a reprecipitation method developed in our laboratory. The C_{60} fine crystals obtained were clearly monodisperse and show interesting various shapes such as spherical, rodlike, fibrous, disk, and octahedral. We were able

Fullerene Nanowhiskers
Edited by Kun'ichi Miyazawa
Copyright © 2012 Pan Stanford Publishing Pte. Ltd.
www.panstanford.com

to selectively control these characteristic sizes and shapes mainly by simply changing the combination of solvents used and the other reprecipitation conditions. In addition, core–shell-type hybridized C_{60} fine crystals composed of C_{60} fine crystals (core) and gold nanoparticles (shell) were also successfully fabricated for the first time. C_{60} fine crystals as a core were prepared by the reprecipitation method using CS_2 as a good solvent and ethanol as a poor solvent. Gold nanoparticles were interestingly deposited on the surface of the C_{60} fine crystals core, without using any reducing reagent, when $HAuCl_4$ aqueous solution was added to 10 mL of C_{60} fine crystals dispersion. The formation of gold nanoparticles is confirmed with scanning electron microscopy, transmission electron microscopy, and powder X-ray diffraction images.

7.1 C_{60} FINE CRYSTALS OF UNIQUE SHAPES AND CONTROLLED SIZE

Many reports[1-4] discussed the unique electronic, optical, and magnetic properties of fullerene molecules. These properties originate in their peculiar conjugated molecular structures, which are spherical and highly symmetrical. Additionally, several interesting papers[5,6] describe bulk fullerene crystals, containing solvent molecules, that are crystallized from solutions. However, only a few studies have been made on fine crystals of fullerenes on a nano-/micrometer scale, besides investigating bulk thin films formed on substrates. The preparation of fullerene fine crystals with a uniform size and shape would permit the control of their specific electronic energy levels and their totally new properties. As a result, we have obtained various unexpected and unique shapes of C_{60} fine crystals with surprisingly monodispersed size.

We first summarize a few previous studies on fullerene fine crystals. Kasai *et al.* have prepared C_{60} nanocrystals of approximately 40–50 nm in size by using the supercritical reprecipitation method, and their optical properties markedly depended on the crystal size.[7] However, only spherical fullerene fine crystals were obtained by using this method. Miyazawa *et al.* have successfully fabricated C_{60} nanowhiskers by the use of the liquid–liquid interfacial precipitation method.[5] In addition, Nakanishi *et al.* have reported the self-assembled nanocones made from chemically modified C_{60} molecules.[8] Moreover, the resulting C_{60} fine crystals were not monodisperse. Recently, Lee and coworkers reported single-crystal C_{60} rods and tubes using cetyltrimethylammonium bromide as a template.[9] In the vapor-driven approach, Bao and coworkers have fabricated hexagonal-shaped C_{60} disks and randomly grown 3D crystals.[10] On the other hand, when Shin *et al.* attempted to prepare C_{60} microdisks by means of the vapor-evaporation solid process on a highly oriented pyrolytic graphite substrate, they could not produce uniform size and shape with a high yield.[11] In general, many experimental difficulties may hinder the control of the size, shape, and monodispersity of C_{60} fine crystals, compared with inorganic fine crystals such as metals or semiconductors.

Given this background, we have further developed a conventional reprecipitation method for the fabrication of C_{60} fine crystals.[12] The advantage of this developed reprecipitation method by skillfully utilizing the intermolecular interaction between the fullerene and good solvent molecules is that the crystal size can be controlled easily by simply changing the experimental conditions such as concentration, the amount of injected solutions, and the temperature of good and poor solvents. Furthermore, we have noticed that bulk C_{60} crystals that were grown from the solutions contain solvent molecules ("crystal solvates") and that their single-crystal surfaces are of considerable high quality.

We have prepared C_{60} fine crystals by using this method and examined the reprecipitation conditions, for example, the concentration and amount of injected solutions, the aging temperature after reprecipitation, and the combination of good and poor solvents. We now report mainly the case of *m*-xylene (good solvent) and 2-propanol (poor solvent).

In the preparation of C_{60} fine crystals by using the reprecipitation method, their size and shape usually depend significantly on the concentration of the injected solutions.[12,13] The overall crystal size clearly reduced with the increasing concentration. Up to the concentration of 2 mM, the size of nanocrystals decreased rapidly from 30 to 1.5 μm and their shape was most fibrous, although some branched structures were also observed. For concentrations over 2 mM, the size changed gradually from 1.5 μm to 100 nm, corresponding to rodlike to spherical shape, and branched structures were not observed.

The amount of injected solutions also affected remarkably the size and shape of C_{60} fine crystals.[12] With a fixed concentration (3 mM), we obtained rodlike nanocrystals at 100 μL injection (Fig. 7.1a) and hollow rodlike nanocrystals at 300 μL injection (Fig. 7.1b). We have again examined this hollow rodlike structure by transmission electron microscopy (TEM) measurement; the hollow structures were present only at the ends of the rod and did not pierce through the rod. For 500 μL and over, we obtained different hollow nanocrystals with a bundle structure (Fig. 7.1c,d). These experimental results were highly reproducible, and the size and shape of C_{60} fine crystals were always the same over several tens of experiments. Moreover, it should be noted that the size distribution in any case was surprisingly monodisperse without exception.

It was found that the aging temperature after reprecipitation also affected the size and shape of the obtained C_{60} fine crystals. After reprecipitation at room temperature (298 K), the resulting dispersion was kept at 273, 313, 333, and 353 K, respectively. C_{60} fine crystals aged at 273 K became shorter rods (Fig. 7.2a), while slightly longer rods were obtained at 313 K (Fig. 7.2b). At 333 and 353 K, the length of rods was further longer than the size of those obtained at 313 K (Fig. 7.2c) while the diameter at both ends of the rods at 353 K was smaller than that at the central part (Fig. 7.2d).

After performing a series of experiments using various combinations of good and poor solvents, we found that the shape of the C_{60} fine crystals obtained is dramatically affected by the choice of the solvent. Figure 7.3 shows many unique C_{60} fine crystals formed in different solvent systems: spherical structure (Fig. 7.3a), long fiber (Fig. 7.3b), flower (Fig. 7.3c,d), star (Fig. 7.3e), plate (Fig. 7.3f), nanoball (Fig. 7.3g), nano-bipyramid (Fig. 7.3h), and microbelt (Fig. 7.3i).

Figure 7.1 SEM images of C_{60} fine crystals prepared by means of the reprecipitation method using *m*-xylene (good solvent) and 2-propanol (poor solvent, 10 mL). The injected amounts of the *m*-xylene solution (3 mM) of C_{60} are (a) 100, (b) 300, (c) 500, and (d) 1000 µL

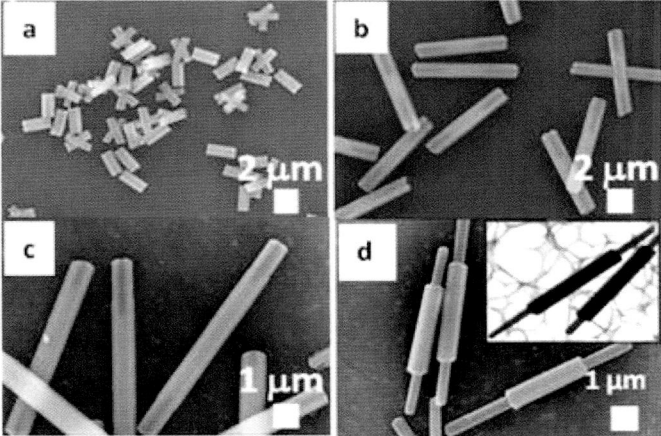

Figure 7.2 SEM images of C_{60} fine crystals prepared at different aging temperatures after reprecipitation at room temperature using *m*-xylene (good solvent) solution (1 mM, 200 µL) injected into 2-propanol (10 mL, poor solvent): (a) 273, (b) 313, (c) 333, and (d) 353 K.

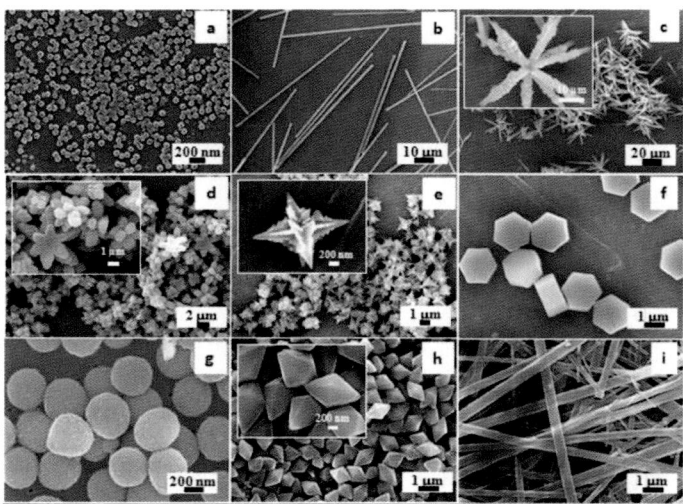

Figure 7.3 Novel and unique shape of C_{60} fine crystals prepared by using the reprecipitation method. (a) Spherical structure formed from m-xylene (good solvent) solution (1 mM, 200 μL) injected into 2-propanol (10 mL, poor solvent) at 353 K and aging at room temperature, (b) long fibers from m-xylene (3 mM, 200 μL, good solvent)/2-butanol (10 mL, poor solvent) with injection and aging at 273 K, (c) flowers from o-xylene (3 mM, 200 μL, good solvent)/2-butanol (10 mL, poor solvent) with injection and aging at room temperature, (d) flowers from p-xylene (0.3 mM, 200 μL, good solvent)/ethanol (10 mL, poor solvent) with injection and aging at room temperature, (e) stars from toluene (1 mM, 500 μL, good solvent)/2-propanol (10 mL, poor solvent) with injection and aging at room temperature, (f) plates from p-xylene (3 mM, 200 μL, good solvent)/1-propanol (10 mL, poor solvent) with injection and aging at room temperature, (g) nanoballs from pyridine (0.3 mM, 200 μL, good solvent)/ethanol (10 mL, poor solvent) with injection and aging at room temperature, (h) nano-bipyramids from CS_2 (0.3 mM, 200 μL, good solvent)/2-propanol (10 mL, poor solvent) with injection and aging at room temperature, and (i) belts from CS_2 (0.3 mM, 600 μL, good solvent)/2-butanol (10 mL, poor solvent) with injection and aging at room temperature.

Next, we describe the internal structure of the various size and shape of C_{60} fine crystals obtained in these experiments. As mentioned previously, C_{60} crystals grown in solutions are "crystal solvates." For instance, a C_{60} bulk crystal fabricated by solvent evaporation from C_{60} m-xylene solutions has the composition C_{60}/m-xylene of 3:2 [14] or 1:1.[6] The C_{60} fine crystals obtained by means of the present reprecipitation method using *m*-xylene (good solvent) and 2-propanol (poor solvent) are also considered to be solvated fine crystals. From Fourier transform infrared spectroscopy, thermogravimetric analysis, and X-ray diffraction (XRD) measurements, we have estimated that the molar ratio of C_{60} to *m*-xylene is approximately 3:2.[12] This ratio was independent of the shape and size, and the crystal structure was hexagonal.

We further consider the formation mechanism of these C_{60} fine crystals. So far, the following two points have been clarified:

(1) C_{60} fine crystals are solvated crystals due to a strong interaction of C_{60} molecules with used good solvents.

(2) The size was almost uniform due to a rapid nucleation.

Taking these phenomena into account in the present reprecipitation method, we speculate that homogeneous nucleation proceeds instantly from the supersaturated state to give monodispersed fine crystals of C_{60}.

We believe this is due to the large difference in the molar ratio of good solvent to poor solvent and to rapid mixing of both solvents after the reprecipitation.

The details of the shape formation process of C_{60} fine crystals, especially in the case of the *m*-xylene and 2-propanol system, are probably as follows. The direction of the crystal growth from C_{60} seed crystals is restricted to the [001] plane, due to its high anisotropy, and consequently the crystal growth proceeds only along one dimension.[15,16] Since the edges of C_{60} seed crystals have a higher surface energy, C_{60} molecules tend to assemble selectively around the edges. As a result, a concentration gradient is formed between the center and edges of the growing C_{60} seed crystal, resulting in a rodlike C_{60} fine crystal. We speculate that the concentration depletion at the central part of the growing seed may lead to rodlike hollowed structures. Such concentration depletion during the growth of inorganic nanoparticles is well known.[16–18] It should also be noted that there exist intermolecular interactions between C_{60} and good solvent molecules. For most molecules, anisotropy of shape or stacking occurs on condensation. However, since fullerenes, especially C_{60} molecules, are spherical and nonanisotropic, the formation of crystals solely from their molecules is very difficult due to the instability of point contacts. In order to avoid this instability, included molecules of good solvents can act as stabilizers or binders. Thus, the nature of the solvent molecules becomes one of the important factors that determine the fine crystal shape. Further investigations will be needed for the detailed elucidation of the shape formation process of C_{60} fine crystals by widely selected combinations of good and poor solvents.

7.2 HYBRIDIZED C_{60} FINE CRYSTALS

Recently, the core–shell-type nanostructures have also attracted much attention.[19–21] In particular, metal nanoshell formed on a core could provide novel optical, electronic, and chemical properties.[22,23] It is most important to choose suitable core and shell materials and to fabricate well-defined interfacial nanostructures in order to control an interfacial interaction. In fact, the enhancement of nonlinear optical properties of hybridized nanomaterials was theoretically predicted[24] and has been actually revealed to occur in the gold nanoparticle dispersed polydiacetylene thin film.[25] We have already prepared the hybridized nanocrystals composed of silver nanoparticles and polydiacetylene nanocrystals by means of the co-reprecipitation method[26] and investigated their hybridized nanostructure and unconventional optical properties. Till date, there are a few reports on metal-coated C_{60} nanocrystals.[27,28] C_{60} derivatives were used in these cases, not C_{60} itself, to fabricate metal–fullerene composites. However, the surface coverage of metal nanoparticles deposited on C_{60} nanocrystals was still very low.

In this section, the fabrication of high-density gold-coated C_{60} fine crystals with unique morphologies of nano-bipyramid and microbelt is described. Nano-bipyramid and microbelt C_{60} fine crystals were also prepared by using the reprecipitation method.[12] A 200 µL of C_{60} solution (2.5 mM), in which the good solvent was CS_2, was quickly injected into vigorously stirred ethanol (10 mL), a poor solvent. The color of the obtained dispersion liquid changed slowly from colorless to light brownish yellow, which indicates the formation of C_{60} fine crystals. The C_{60} fine crystal dispersion liquid should be further allowed to stand for several hours to completely finish nanocrystallization. The typical scanning electron microscopy (SEM) images of the resulting nano-bipyramid and microbelt C_{60} fine crystals are exhibited in Fig. 7.4. The nano-bipyramid C_{60} fine crystals are monodispersed 12-faced body in shape, which have a hexagonal symmetry as shown in Fig. 7.4a and have an average length of 2.5 μm in the long axis. The nano-bipyramid C_{60} fine crystals have the hexagonal closed packed (hcp) crystal structure confirmed by powder XRD measurement. Figure 7.4b displays a TEM image of bipyramid C_{60} fine crystals. The inset in Fig. 7.4b shows the electron diffraction (ED) pattern, which is indexed with the [011] orientation. On the other hand, a C_{60} nanobelt was similarly prepared from the same solvent system by only increasing the injected amount of C_{60} solution from 200 to 500 μL. The average cross-sectional area of the C_{60} nanobelt is approximately 40 × 200 nm, and the length is greater than 10 μm (Fig. 7.4c). Crystal structure analysis suggested that the nanobelt has a mixed phase of face-centered cubic (fcc) structure and monoclinic structure. As shown in Fig. 7.4d, the TEM image may suggest that the nanobelt is polycrystalline, according to the structure analysis result. The ED pattern of the nanobelt was indexed with the direction [110] assuming the fcc symmetry.

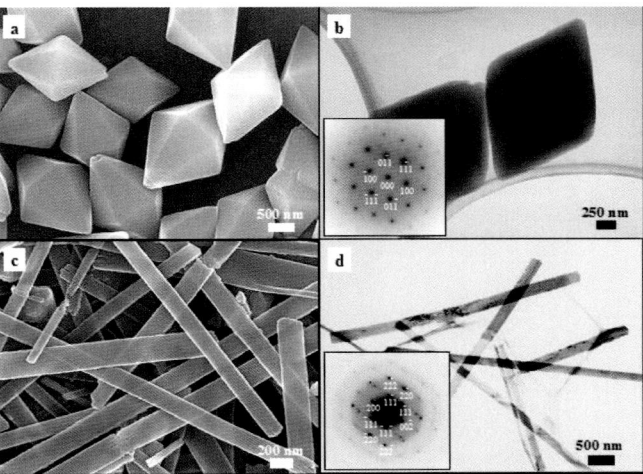

Figure 7.4 (a) SEM and (b) TEM images of bipyramid C_{60} fine crystals. (c) SEM and (d) TEM images of C_{60} nanobelts. The insets exhibit the ED patterns of C_{60} fine crystals.

Afterward, 80 mL of the resulting bipyramid fine crystals dispersion was concentrated to 10 mL and then 500 μL of $HAuCl_4$ aqueous solution (22.2 mM) was added. The mixed dispersion was slowly stirred for 2 h at 40°C. Fig. 7.5a displays gold-coated C_{60} fine crystals with nano-bipyramid morphology. The average size of the deposited gold nanoparticles is in the range of 10–20 nm. The coverage of gold nanoparticles deposited on the C_{60} core is much higher than that in the previous research.[28] The TEM image of gold-coated C_{60} fine crystals confirms the high surface coverage as shown in Fig. 7.5b. The ED pattern shown in the inset of Fig. 7.5b displays the diffraction rings from Au (111), (200), (220), (311), and (222). Figure 7.5c,d exhibits SEM and TEM images of high-density gold-coated C_{60} nanobelts obtained by using a similar treatment. The ED pattern shown in the inset of Fig. 7.5d also confirms the formation of gold nanoparticles. These results indicate that the deposition of gold nanoparticles on C_{60} fine crystal cores proceeded, being independent of their unique morphologies. From the SEM and TEM images in Fig. 7.5, it is evident that few isolated gold nanoparticles coexisted with gold-coated C_{60} fine crystals.

Figure 7.5 (a) SEM and (b) TEM images of gold-coated bipyramid C_{60} fine crystals. (c) SEM and (d) TEM images of gold-coated C_{60} nanobelts. The insets in (b) and (d) show the ED patterns of gold-coated C_{60} fine crystals.

Figure 7.6 shows the powder XRD patterns of C_{60} and gold-coated C_{60} fine crystals. From these XRD patterns, we could also confirm that C_{60} fine crystals have the same hexagonal phase as C_{60} bulk crystal (hcp structure, with c/a = 1.633).[29] For example, one can clearly see the XRD diffraction peaks of Au (111), (200), (220), and (311) in Fig. 7.6.

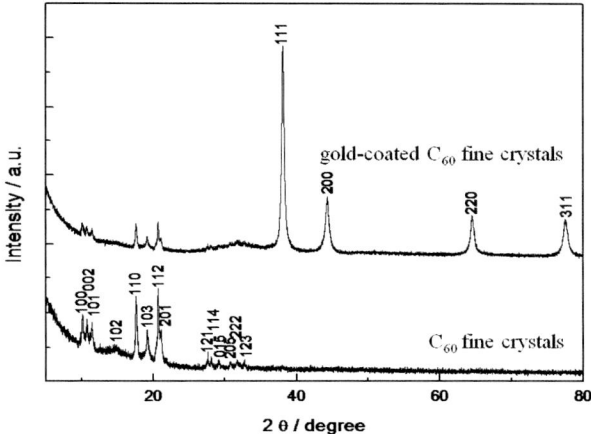

Figure 7.6 Powder XRD patterns of C_{60} fine crystals and gold-coated C_{60} fine crystals.

7.3 CONCLUSION

To the best of our knowledge, the reprecipitation method described here is the simplest and most convenient of all the methods developed so far to fabricate C_{60} fine crystals, and this chapter represents the first discovery concerning the vast sizes and shapes of C_{60} fine crystals. In addition, we have succeeded in the fabrication of gold-coated C_{60} fine crystals only by the addition of $HAuCl_4$ followed by the subsequent heat treatment. Gold-coated C_{60} fine crystals are expected to have great potential in applications such as optoelectronics, advanced catalysis, bio/chemical sensors, and third-order nonlinear optics.

Acknowledgments

This work was financially supported by a Grant-in-Aid for Young Scientists B (17710086) from the Ministry of Education, Culture, Sports, Science and Technology, the Exploratory Research Program for Young Scientists, Tohoku University, a research grant from the Hosokawa Powder Technology Foundation, IMRAM Project, Tohoku University, and the 2007 Tohoku University Global COE program "International Center of Research & Education for Molecular Complex Chemistry" for young scientists.

References

1. W. Andreoni, *The Physics of Fullerene-Based and Fullerene-Related Materials Series, Physics and Chemistry of Materials with Low-Dimensional Structures*, Vol. 23, Kluwer Academic, Dordrecht, 2000.
2. M. Akada, T. Hirai, J. Takeuchi, T. Yamamoto, R. Kumashiro, and K. Tanigaki, *Phys. Rev. B*, **73**, 094509 (2006).
3. H. Ohashi, K. Tanigaki, R. Kumashiro, S. Sugihara, S. Hiroshiba, S. Kimura, K. Kato, and M. Takata, *Appl. Phys. Lett.*, **84**, 520 (2004).
4. F. Yang and S. R. Forrest, *Adv. Mater.*, **18**, 2018 (2006).
5. K. Miyazawa, Y. Kuwasaki, A. Obayashi, and M. Kuwabara, *J. Mater. Res.*, **17** 83 (2002).
6. L. Wang, B. Liu, S. Yu, M. Yao, D. Liu, Y. Hou, T. Cui, G. Zou, B. Sundqvist, H. You, D. Zhang, and D. Ma, *Chem. Mater.*, **18**, 4190 (2006).
7. H. Kasai, S. Okazaki, T. Hanada, S. Okada, H. Oikawa, T. Adschiri, K. Arai, K. Yase, and H. Nakanishi, *Chem. Lett.*, 139 (2000).
8. T. Nakanishi, W. Schmitt, T. Michinobu, D. G. Kurth, and K. Ariga, *Chem. Commun.*, 5982 (2005).
9. H. X. Ji, J. S. Hu, Q. X. Tang, W. G. Song, C. R. Wang, W. P. Hu, L. J. Wan, and S. T. Lee, *J. Phys. Chem. C*, **111**(28), 10498 (2007).
10. A. L. Briseno, S. C. B. Mannsfeld, M. M. Ling, S. Liu, R. J. Tseng, C. Reese, M. E. Roberts, Y. Yang, F. Wudl, and Z. Bao, *Nature*, **444**, 913 (2006).
11. H. S. Shin, S. M. Yoon, Q. Tang, B. Chon, T. Joo, and H. C. Choi, *Angew. Chem. Int. Ed.*, **47**, 693 (2008).

12. A. Masuhara, Z. Tan, H. Kasai, H. Nakanishi, and H. Oikawa, *Jpn. J. Appl. Phys.*, **48**, 050206 (2009).
13. Z. Tan, A. Masuhara, H. Kasai, H. Nakanishi, and H. Oikawa, *Jpn. J. Appl. Phys.*, **47**, 1426 (2008).
14. M. V. Korobov, E. B. Stukalin, A. L. Mirakyan, I. S. Neretin, Y. L. Slovokhotov, A. V. Dzyabchenko, A. I. Ancharov, and B. P. Torochko, *Carbon*, **41**, 2743 (2003).
15. R. Alargova, G. S. Deguchi, and K. Tsuji, *J. Am. Chem. Soc.*, **123**, 10460 (2001).
16. B. Mayers and Y. N. Xia, *Adv. Mater.*, **14**, 279 (2002).
17. E. Matijevic, *Langmuir*, **10**, 8 (1994).
18. J. Aizenberg, A. J. Black, and G. M. Whitesides, *Proc. Am. Chem. Soc., PMSE*, **81**, 2 (1999).
19. S. L. Westcott, S. J. Oldenburg, T. R. Lee, and N. J. Halas, *Langmuir*, **14**, 5396, (1998).
20. V. G. Pol, A. Gedankan, and J. Caldenron-Moreno, *Chem. Mater.*, **15**, 1111 (2003).
21. W. L. Shi, Y. Sahoo, M. T. Swihart, and P. N. Prasad, *Langmuir*, **21**, 1610 (2005).
22. S. J. Oldenburg, G. D. Hale, J. B. Jackson, and N. J. Halas, *Appl. Phys. Lett.*, **75**, 1063 (1999).
23. J. N. Gillet and M. J. Meunier, *Phys. Chem. B*, **109**, 8733 (2005).
24. A. E. Neeves and M. H. Birnboin, *J. Opt. Soc. Am. B*, **6**, 787 (1989).
25. A. W. Olsen and Z. H. Kafafi, *J. Am. Chem. Soc.*, **113**, 7758 (1991).
26. A. Masuhara, H. Kasai, S. Okada, H. Oikawa, M. Terauchi, M. Tanaka, and H. Nakanishi, *Jpn. J. Appl. Phys.*, **40**, L1129 (2001).
27. N. Sun, Y. X. Wang, Y. L. Song, Z. X. Guo, L. M. Dai, and D. B. Zhu, *Chem. Phys. Lett.*, **344**, 277 (2001).
28. P. Zhang, J. X. Li, D. F. Liu, Y. J. Qin, Z. X. Guo, and D. B. Zhu, *Langmuir*, **20**, 1466 (2004).
29. E. V. Skokan, I. V. Arkhangelskii, D. E. Izotov, N. V. Chelovskaya, M. M. Nikulin, and Y. A. Velikodnyi, *Carbon*, **46**, 803 (2005).

Chapter 8

IN SITU TRANSMISSION ELECTRON MICROSCOPY OF FULLERENE NANOWHISKERS AND RELATED CARBON NANOMATERIALS

Tokushi Kizuka

Institute of Materials Science, University of Tsukuba,1-1-1, Tennoudai, Tsukuba, Ibaraki 305-8573, Japan
kizuka@ims.tsukuba.ac.jp

We present the structure and electrical, mechanical, and optical properties of isolated crystalline fullerene nanowhiskers and carbon nanocapsules investigated by in situ high-resolution transmission electron microscopy. We first explain the methodology, followed by examples of in situ observation for each subject, i.e., elastic properties of C_{60} nanowhiskers and the formation, luminescence, conductance, and deformation of carbon nanocapsules.

8.1 INTRODUCTION

The synthesis of a new form of C_{60} single crystals with high aspect ratios of length to diameter, i.e., fullerene nanowhiskers (FNWs), by Miyazawa et al.[1,2] has led to the study of the structure and properties of FNWs for their application in advanced nanometer-scale functional and structural devices. To analyze their mechanical properties, manipulation of individual FNWs for both deformation and nanonewton-scale force measurements is required. For investigating their electrical properties, it is necessary to attach at least two electrodes on both ends of individual FNWs. Although such operations are challenging even with the use of advanced technologies, some results have been obtained for FNWs and related carbon nanomaterials by in situ high-resolution transmission electron microscopy (TEM) in which conductance and force are measured.[3-6] In this chapter, first in situ TEM is described and then some results are presented.

8.2 IN SITU TEM IN THE STUDY OF NANOMATERIALS

The described in situ TEM was developed on the basis of in situ high-resolution TEM combined with subnanonewton force measurements used in atomic force microscopy (AFM) and with electronic conductance measurements used in scanning tunneling microscopy (Fig. 8.1).[4,7] First, we prepare nanometer-sized tips of noble metals as electrodes: a noble metal is deposited on a Si cantilever with a nanotip by vacuum deposition. The cantilever tip is attached to the front of a cylindrical piezoelement on a cantilever holder for TEM. Then, a noble metal plate of 0.2 mm thickness is attached to the second plate holder. The contact edge of the plate is thinned to 5–20 nm by argon ion milling. On the plate, FNWs or carbon nanomaterials are dispersed. The cantilever and plate holders are then inserted into the in situ TEM, as performed at the University of Tsukuba. The specimen chamber of the microscope is evacuated, first by a turbomolecular pump and then by an ion pump, resulting in a vacuum of 1×10^{-5} Pa. Inside the microscope, the cantilever tip is brought into contact with an FNW or a carbon nanomaterial on the edge surface of the opposing plate by piezomanipulation. The bias voltage is applied between the tip and the plate to measure conductance. A series of these manipulations are performed at room temperature. The structural dynamics during the process is observed in situ by lattice imaging via high-resolution TEM using a television capture system. The time resolution of image observation is 17 ms. The force applied between the tip and the plate is simultaneously measured by optical detection of the cantilever deflection. The spring constant of the cantilever is typically ~5 N/m. The electrical conductance is measured by a two-terminal method. The results of high-resolution imaging and signal detection in this system are simultaneously recorded and analyzed for each image.

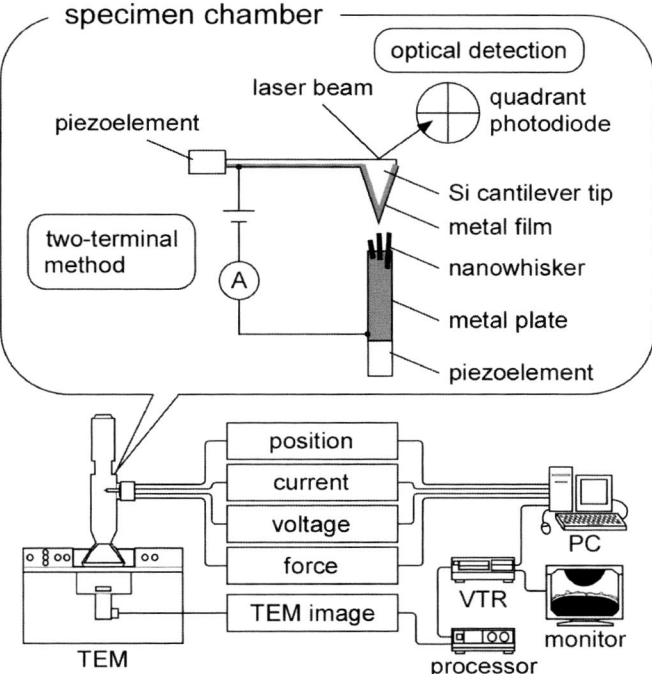

Figure 8.1 Illustration of in situ transmission electron microscopy for observing the structures and electrical conductance of nanowhiskers and the force acting on them.

8.3 EXAMPLES OF MEASUREMENTS

8.3.1 Elastic Properties of FNWs

After FNWs are dispersed on the plate, they often protrude from the plate edge; one end of such FNWs is free in the vacuum and another is fixed on the plate. It is easy to compress these FNWs along their long axes using the cantilever tip.[8,9] Thus, we first demonstrate the buckling of FNWs (Fig. 8.2). In response to compression, initially the FNWs bend elastically and then buckle. During this process, the force acting on the FNWs is measured by optical deflection, and its variation is analyzed as a function of tip displacement, i.e., compression depth (Fig. 8.3). The Young's modulus of the FNWs is estimated using Euler's formula. The buckling force $P_{buckling}$ of a material with a columnar shape is given by

$$P_{buckling} = k\frac{\pi^2 EI}{L^2}$$

where k is a fixity coefficient, E is Young's modulus, I is the geometrical moment of inertia, and L is the length of the column.[9] Here, k is selected such that it corresponds to the fixed-free end condition, i.e., 0.25. Since FNWs are columnar, I is given by

$$I = \frac{\pi d^4}{64}$$

where d is the diameter of the FNW.

When an FNW is supported by two fine side edges of the plate, the edges play the role of fulcra. Thus, a bending test can be performed in this situation. The bending force $P_{bending}$ of a material with a columnar shape is given by

$$P_{bending} = \frac{3EI\ell}{x^2(\ell-x)^2}y$$

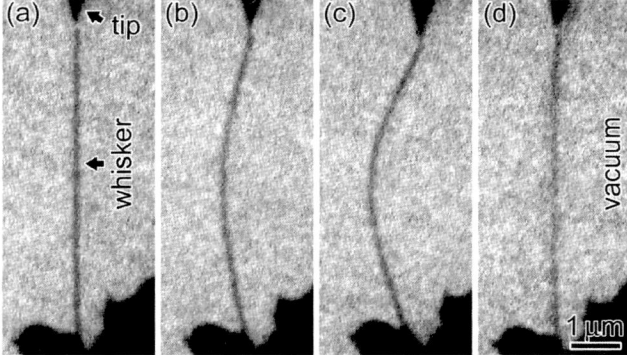

Figure 8.2 Time-sequential images of the buckling process of a C_{60} whisker. The tip of the cantilever and the edge surface of the microgrid can be observed at the top and the bottom of the image, respectively. The bright regions represent the vacuum. (Reprinted with permission from Asaka et al.,[9] © 2006, American Institute of Physics.)

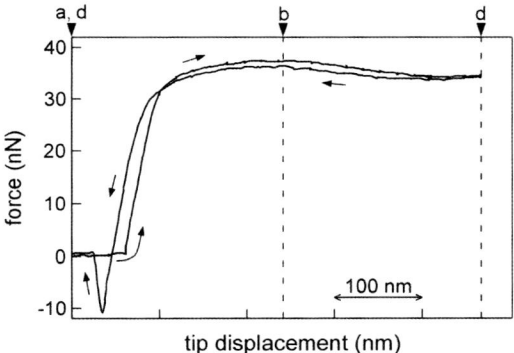

Figure 8.3 Variation in force during buckling presented as a function of cantilever tip displacement. States a–d correspond to images (a)–(d) in Fig. 8.2. (Reprinted with permission from Asaka et al.,[9] © 2006, American Institute of Physics.)

where ℓ is the effective deformation length, χ is the length from the nearest fulcrum to the point at which the force is applied, and y is the flexure of the simple columnar beam.[10] The structural parameters d, L, ℓ, χ, and y are estimated from TEM images. P_{buckling} and P_{bending} are directly measured by force detection. Therefore, Young's modulus is calculated both in the buckling and bending tests and is plotted against the outer diameter (Fig. 8.4).

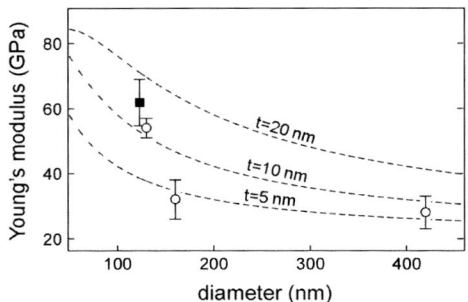

Figure 8.4 Young's modulus of C_{60} nanowhiskers plotted against the outer diameter. The moduli obtained by the bending test and the buckling test are represented by solid squares and open circles, respectively. The dashed lines show the calculated Young's modulus on the basis of the core–shell model for nanowhiskers with thicknesses t = 5, 10, and 20 nm. (Reprinted with permission from Saito et al.,[10] © 2009, The Japan Society of Applied Physics.)

Young's modulus increases as the outer diameter decreases. This leads to the presumption that the elastic property of the side surfaces is emphasized in the total Young's modulus owing to size reduction. It is proposed that C_{60} nanowhiskers have core–shell structures: the structure, composition, and bonding of the region near the side surfaces differ from those of the inner region along the long axis.[11] Assuming that the Young's moduli of the inner and outer regions of an NW differ, the total Young's modulus is expressed by the following formula:

$$E = E_s + (E_c - E_s)\left(1 - \frac{2t}{d}\right)^4 \left(t < \frac{d}{2}\right),$$

where E_s and E_c are the Young's moduli of the outer and inner regions, respectively, d is the outer diameter of the whisker, and t is the thickness of the outer region.[10] For E_c, the Young's modulus of large C_{60} crystals, i.e., ~20 GPa, can be used, as the contribution of the surface region to the total Young's modulus is small.[11,12] For E_s, first we need to consider the thickness of the outer region to be 10–20 nm. We may assume the average Young's modulus of the C_{60} nanotubes, i.e., 84.5 GPa, as that of the outer region.[13] Using these values, the total Young's modulus is calculated. The Young's moduli of FNWs with t = 5, 10, and 20 nm closely fit the measured values, particularly when t is 10 nm (Fig. 8.4). Therefore, these results support the core–shell structure model for FNWs.

The bending of crystalline fullerene nanotubes and the fracture of FNWs can also be investigated by this in situ TEM.[9, 13]

8.3.2 Carbon Nanocapsules

8.3.2.1 Formation

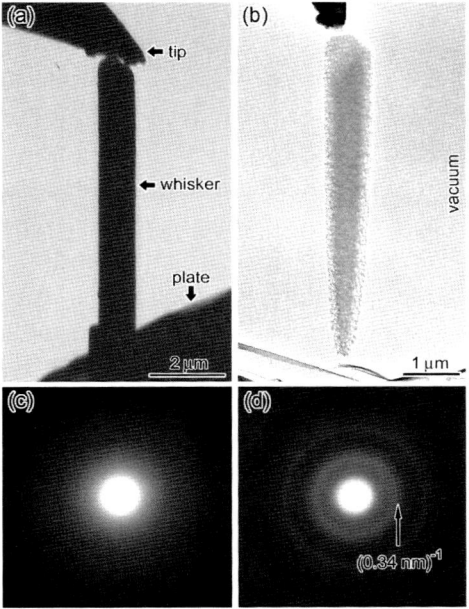

Figure 8.5 Bright-field images of (a) amorphous carbon whisker heated at 1373 K before passing current; (b) the same whisker after application of a voltage of 4 V. The whisker protrudes from a perforated silver plate in high vacuum. The bias voltage was applied after the whisker was sandwiched by the cantilever tip and the plate. (c) and (d) are selected-area electron diffraction patterns from the whisker observed in (a) and (b), respectively. (Reprinted with permission from Asaka et al.,[17] © 2006, American Institute of Physics.)

The electrical resistivity of FNWs ranges from 10^6 to 10^8 Ω m, similar to that of crystalline C_{60} films and plates.[14] By heating at 1373 K in vacuum, the structure of FNWs transforms into an amorphous form, and the resistivity becomes lower than that of pristine FNWs, i.e., 4×10^{-4} Ω m.[15] This reduced resistivity allows us to heat the amorphous carbon whiskers (ACWs) by passing current at a bias voltage of several volts, resulting in a synthesis of hollow multiwalled carbon nanocapsules (CNCs), which are multiwalled fullerenes (Figs. 8.5 and 8.6).[16–19]

8.3.2.2 Luminescence

After the formation of CNCs by passing current, the whiskers remain long and thin (Fig. 8.6). We can then connect the whiskers again with the electrode. When the bias voltage is increased from 0 V, the whiskers start to illuminate (Fig. 8.7).[17] The emission intensity increases with an increase in the electric power from a threshold power of 0.84 mV. The power applied at the whisker is lower than that required for high-intensity electroluminescence in other organic light-emitting materials, e.g., aluminum 8-hydroxyquinoline complex.[20] As the electric power increases, the full width at half maximum of the spectra reduces: the width is ~240 nm at 2.6 mW. At this electric power, the luminescence is easily seen by the naked eye at a distance of 250 mm. The maximum peak ranges from 1.70 to 1.77 eV, corresponding to a wavelength of 700–730 nm. Most of the emission intensity is caused by electroluminescence. These peaks are similar to those of photoluminescence from the C_{60} crystals (~730 nm).[21–25] Thus, luminescent filaments can be easily synthesized from FNWs.

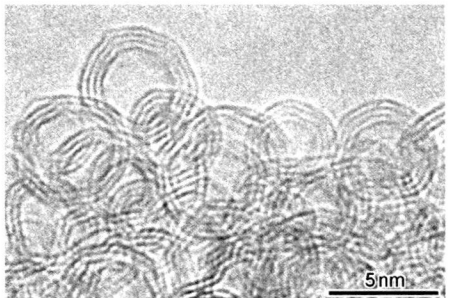

Figure 8.6 High-resolution image of carbon nanocapsules on whisker surface after application of 4 V bias voltage.

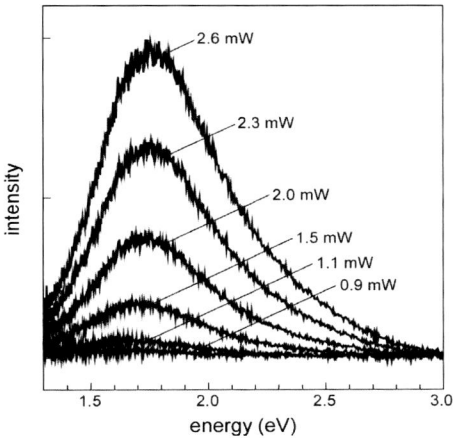

Figure 8.7 Emission spectra of a whisker composed of carbon capsules at various electric powers. (Reprinted with permission from Asaka et al.,[17] © 2006, American Institute of Physics.)

8.3.2.3 Conductance

One of the aggregated CNCs on the whisker surfaces can be picked up and sandwiched between the two metallic nanotips by piezomanipulation.[26,27] This structure corresponds to a type of single-molecule junction (Fig. 8.8).[28–30] The conductance can then be measured, and the relationship to the junction structures can be investigated (Fig. 8.9).[27]

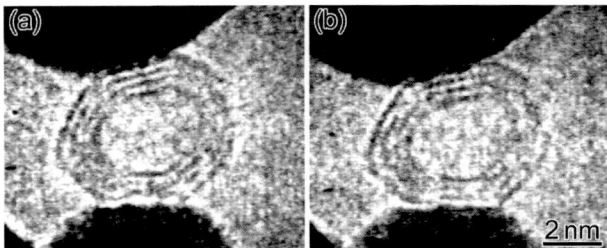

Figure 8.8 Time-sequential high-resolution images of a junction of a single multiwalled carbon nanocapsule during compression. The dark regions at the top and bottom show the gold cantilever tip and the edge of a gold plate, respectively. The bright region is the vacuum. (Reprinted with permission from Asaka et al.,[27] © 2007, The American Physical Society.)

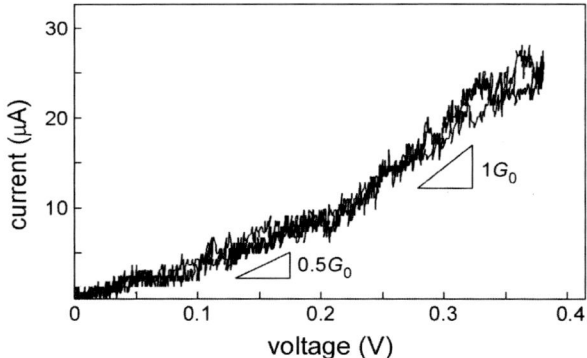

Figure 8.9 Variations in current during the compression presented in Fig. 8.8 under a bias voltage of 0.29 V. The conductance increased from $0.5G_0$ to $1G_0$ owing to compression. (Reprinted with permission from Asaka et al.,[27] © 2007, The American Physical Society.)

The differential conductance of the junction at bias voltages from 0 to 0.2 V corresponds to half the quantized conductances, i.e., $0.5G_0$ (where $G_0 = 2e^2/h$ is the conductance quantum, e is the electron charge, and h is Planck's constant). At higher bias voltages, an electrostatic attractive force acts between the two metallic nanotips; the interface structure between the CNC and the nanotip changes. As a result, the differential conductance increases to $1G_0$. In contrast, when the total number of wall layers decreases owing to the breakdown of the outermost layer, the differential conductance decreases discontinuously from $1G_0$ to $0.5G_0$. At this discontinuous reduction, the maximum current decreases by 19 μA. Thus, the electron transport is related to several layers of the CNCs, similar to that through carbon nanotubes when their surface is in contact with electrodes.

The conductance of a CNC is comparable to the quantized levels defined by G_0. CNC junctions have higher structural stability under voltage application than metallic nanocontacts, which show regular quantized conductance.[6,31–34]

The maximum current and the conductance of CNC junctions can be controlled by selecting the number of wall layers, which is expected to lead to developments in the application of CNCs in carbon device technologies.

8.3.2.4 Mechanical properties

The deformation of isolated CNCs can also be observed.[26] The mechanical properties of CNCs are investigated on an atomic scale. For example, a CNC with diameter 2.50 ± 0.06 nm is compressed using a gold nanotip until a strain of 0.06 is achieved at a force of up to 4.5 ± 0.5 nN. Through this compression, the CNC adhered to the gold tip. Thus, CNCs can be assembled with metal elements by mechanical processing. At a higher load, CNCs collapse entirely. However, after subsequent release of the load, the CNCs regain their initial structures. Thus, CNCs possess a high resistance to compression, which is sufficient for application in structural and functional nanodevices.

8.4 CONCLUSIONS AND OUTLOOK

The structure and electrical, mechanical, and optical properties of FNWs can be investigated on the basis of individual observation and measurements. In situ high-resolution TEM allows us to perform such experiments. CNCs were successfully synthesized by using the electrical contact method in a series of experiments.

In the electrical contact method, we can select the position of voltage application by piezomanipulation of a nanometer-sized electrode. As a result, we can locally heat an ACW and synthesize CNCs. The application of our method to amorphous carbon nanodots is expected to result in a similar localized formation of CNCs. The conductance of a CNC is comparable to quantized conductance. Thus, our method leads to the configuration control of carbon quantum dot arrays on electronic substrates. The research demonstrated in this chapter provides a basis for the development of FNWs and related carbon nanomaterials including CNCs in nanocarbon device technology.

Acknowledgments

I acknowledge the participation of Dr. Ryoei Kato, Dr. Kazuma Saito, Dr. Koji Asaka, and Dr. Kun'ichi Miyazawa as coauthors in the referenced work.

References

1. K. Miyazawa, A. Obayashi, and M. Kuwabara, *J. Am. Ceram. Soc.*, **84**, 3037 (2001).
2. K. Miyazawa, Y. Kuwasaki, A. Obayashi, and M. Kuwabara, *J. Mater. Res.*, **17**, 83 (2002).
3. T. Kizuka, K. Yamada, S. Deguchi, M. Naruse, and N. Tanaka, *Phys. Rev. B*, **55**, R7398 (1997).
4. T. Kizuka, H. Ohmi, T. Sumi, K. Kumazawa, S. Deguchi, M. Naruse, S. Fujisawa, S. Sasaki, A. Yabe, and Y. Enomoto, *Jpn. J. Appl. Phys.*, **40**, L170 (2001).
5. T. Kizuka, Y. Takatani, K. Asaka, and R. Yoshizaki, *Phys. Rev. B*, **72**, 035333 (2005).
6. T. Kizuka, *Phys. Rev. B*, **77**, 155401 (2008).
7. T. Kizuka and N. Tanaka, *Phil. Mag. Lett.*, **69**, 135 (1994).
8. R. Kato, K. Asaka, K. Miyazawa, and T. Kizuka, *Jpn. J. Appl. Phys.*, **45**, 8024 (2006).
9. K. Asaka, R. Kato, K. Miyazawa, and T. Kizuka, *Appl. Phys. Lett.*, **89**, 071912 (2006).

10. K. Saito, K. Miyazawa, and T. Kizuka, *Jpn. J. Appl. Phys.*, **48**, 010217 (2009).

11. K. Miyazawa, J. Minato, T. Yoshii, M. Fujno, and T. Suga, *J. Mater. Res.*, **20**, 688 (2005).

12. S. Hoen, N. G. Chopra, X. D. Xiang, R. Mostovoy, J. Hou, W. A. Vareka, and A. Zettl, *Phys. Rev. B*, **46**, 12737 (1992).

13. T. Kizuka, K. Saito, and K. Miyazawa, *Diam. Relat. Mater.*, **17**, 972 (2008).

14. K. Miyazawa, Y. Kuwasaki, K. Hamamoto, S. Nagata, A. Obayashi, and M. Kuwabara, *Surf. Interface Anal.*, **35**, 117 (2003).

15. K. Miyazawa, J. Minato, H. Zhou, T. Taguchi, I. Honma, and T. Suga, *J. Eur. Ceram. Soc.*, **26**, 429 (2006).

16. K. Asaka, R. Kato, R. Yoshizaki, K. Miyazawa, and T. Kizuka, *Phys. Rev. B*, **76**, 113404 (2007).

17. K. Asaka, R. Kato, Y. Maezono, R. Yoshizaki, K. Miyazawa, and T. Kizuka, *Appl. Phys. Lett.*, **88**, 051914 (2006).

18. T. Kizuka, R. Kato, and K. Miyazawa, *Carbon*, **47**, 138 (2009).

19. T. Kizuka, R. Kato, and K. Miyazawa, *Nanotechology*, **20**, 105205 (2009).

20. J. Kalinowski, *J. Phys. D*, **32**, R179 (1999).

21. C. Reber, L. Yee, J. Mckiernan, J. I. Zink, R. S. Williams, W. M. Tong, D. A. A. Ohlberg, R. L. Whetten, and F. Diederich, *J. Phys. Chem.*, **95**, 2127 (1991).

22. J. Feldmann, R. Fischer, W. Guss, E. O. Göbel, S. Schmitt-Rink, and W. Krätschmer, *Europhys. Lett.*, **20**, 553 (1992).

23. M. Matus, H. Kuzmany, and E. Sohmen, *Phys. Rev. Lett.*, **68**, 2822 (1992).

24. W. Guss, J. Feldmann, E. O. Göbel, C. Taliani, H. Mohn, W. Müller, P. Häussler, and H.-U. T. Meer, *Phys. Rev. Lett.*, **72**, 2644 (1994).

25. Y. Wang, J. M. Holden, A. M. Rao, P. C. Eklund, U. D. Venkateswaran, D. Eastwood, R. L. Lidberg, G. Dresselhaus, and M. S. Dresselhaus, *Phys. Rev. B*, **51**, 4547 (1995).

26. K. Asaka, R. Kato, K. Miyazawa, and T. Kizuka, *Appl. Phys. Lett.*, **89**, 191914 (2006).

27. K. Asaka, R. Kato, R. Yoshizaki, K. Miyazawa, and T. Kizuka, *Phys. Rev. B*, **76**, 113404 (2007).
28. J. K. Gimzewski and C. Joachim, *Science*, **283**, 1683 (1999).
29. C. Dekker, *Phys. Today*, **52**, 22 (1999).
30. C. Joachim, J. K. Gimzewski, and A. Aviram, *Nature (London)*, **408**, 541 (2000).
31. N. Agraït, J. G. Rodrigo, and S. Vieira, *Phys. Rev. B*, **47**, 12345 (1993).
32. J. I. Pascual, J. Méndez, J. Gómez-Herrero, A. M. Baró, N. García, and V. T. Binh, *Phys. Rev. Lett.*, **71**, 1852 (1993).
33. J. M. Krans, J. M. Van Ruitenbeek, V. V. Fisun, I. K. Yanson, and L. J. De Jongh, *Nature (London)*, **375**, 767 (1995).
34. H. Yasuda and A. Sakai, *Phys. Rev. B*, **56**, 1069 (1997).

Chapter 9

MECHANICAL BEND TESTING OF FULLERENE NANOWHISKERS

Stepan Lucyszyn and Michael P. Larsson

Department of Electrical and Electronic Engineering, Imperial College London,
London SW7 2AZ, United Kingdom
s.lucyszyn@imperial.ac.uk

Little has been published on the mechanical bend testing of fullerene (nano)whiskers due to the inherent difficulties in physically mounting such small test samples. Earlier reported results suggested the Young's modulus values of 32 and 54 GPa for 130 and 160 nm diameter C_{60} whiskers, respectively, using compressive deformation techniques. An experimental bespoke silicon-based microelectromechanical system has also been developed to extract another value. It has been found, through parameter extraction techniques, that a Young's modulus of only ~2 GPa is obtained with a C_{60} whisker having a diameter of 4 µm. By combining these three data points, there is now evidence to suggest an inverse proportionality relationship between the Young's modulus and the diameter of a C_{60} whisker.

Fullerene Nanowhiskers
Edited by Kun'ichi Miyazawa
Copyright © 2012 Pan Stanford Publishing Pte. Ltd.
www.panstanford.com

9.1 INTRODUCTION

Since the discovery of fullerene C_{60}, by Kroto et al.,[1] advances in the field of fullerene chemistry have led to a number of interesting developments, such as the carbon nanotube (CNT). The C_{60} molecule consists of carbon atoms located at the nodes of a series of hexagons and pentagons arranged in a cage lattice, defined by alternating double and single bonds. In contrast to CNTs, which can be visualized as rolled-up sheets of graphene, C_{60} whiskers contain no long-range hollow structure. Instead, structural characterization has indicated the whiskers are solid, consisting of a series of C_{60} molecules, bound through a combination of van der Waals interactions and chemical bonds. The latter suggests that C_{60} whiskers contain chains of polymerized C_{60} within their structure. A transition from a poly- to single-crystalline structure may occur when the diameter of C_{60} whiskers decreases to approximately 1 μm.[2,3] The absence of grain boundaries, combined with enhanced intermolecular bonding, are the main reasons why C_{60} nanowhiskers are expected to exhibit superior electrical and mechanical properties when compared to those of their larger diameter counterparts.

In recent years, work at the University of Tsukuba and National Institute for Materials Science (both in Tsukuba, Japan) has led the way in determining the mechanical properties of C_{60} nanowhiskers using compressive deformation techniques.[4] Asaka et al. determined the Young's modulus values of 32 and 54 GPa for 130 and 160 nm diameter C_{60} whiskers, respectively. This is in excess of the 8.3–20 GPa range that they reported in their survey for bulk C_{60} crystals.[4]

More recently, a bespoke microelectromechanical systems (MEMS) electrothermal 4-point bend tester was developed at Imperial College London for the mechanical characterization of C_{60} (nano)whiskers for operation under an optical microscope.[5]

C_{60} molecules are known to polymerize under high temperature and high pressure conditions[2,3] and in the presence of ultraviolet radiation. Relatively recently, Miyazawa et al. reported a technique for growing needlelike structures of C_{60} from a solution at room temperature.[2,3] Here, the needles of C_{60} can grow in arbitrary directions through a process of liquid–liquid interfacial precipitation

(LLIP) between C_{60} saturated toluene and isopropyl alcohol (IPA). Diameters ranged from hundreds of nanometers to several microns, with lengths reaching several hundreds of microns. Needles with diameters less than 1 µm are referred to as C_{60} nanowhiskers, and those with larger diameters are referred to as C_{60} whiskers.

9.2 MECHANICAL BEND TESTING

Techniques for determining the mechanical properties of CNTs, in particular, have been widely reported in the open literature. The most common involves tensile loading using the tip of an atomic-force microscope (AFM) probe.[6] In such situations, either the CNTs are attached to the AFM tip by pressure contact or are grown on. The former is more common; nevertheless, the nonideal attachment between the CNT and the substrate and the tip of the AFM limit the accuracy of the data obtained. The problem of gripping can be avoided by reversing the loading direction of the AFM tip and performing a buckling test.

Very little has been published in the open literature on the mechanical characterization of C_{60} nanowhiskers. However, relatively recently, the Young's modulus (E) for a C_{60} nanowhisker was ascertained from a buckling test in a transmission electron microscope (TEM), using a probe tip with position control and force feedback.[4] Unfortunately, with this approach, the point of loading shifts during the test.

Problems with sample gripping/loading during mechanical testing can be resolved through the use of bend tests. Such tests are popular in mechanical engineering for the characterization of brittle materials. In such cases, 3- and 4-point tests are employed by convention, and the deflection and fracture data can be used to determine E and fracture strength, which for brittle materials is known as the modulus of rupture. Indeed, micromachined tensile testers have successfully been developed for the characterization of materials commonly used in MEMS technology, such as silicon and polysilicon. These devices are quite sophisticated, incorporating either electrothermal or electrostatic actuation. Force feedback can be achieved using electrostatic comb-drive sensors. It is the successful application of such devices that is the inspiration behind the mechanical testing of C_{60} (nano)whiskers.[5] As C_{60} whiskers are large enough to be observed under an optical microscope, this lends itself well to the use of simplified MEMS mechanical testers to evaluate their deflection and fracture characteristics.

The tensile loading method for CNTs, the bucking test method for C_{60} nanowhiskers, and the MEMS bend tester for C_{60} (nano) whiskers will now be described in more detail.

9.3 TENSILE LOADING

Single-wall carbon nanotube (SWCNT) ropes have been proven to have a very high breaking strength.[6] As a result, there is considerable interest in its applications for creating very strong and lightweight materials.

SWCNTs have been made using the laser ablation method and purified by reflux and filtration.[6] The resulting SWCNT "paper" was torn apart to make individual SWCNT ropes that protrude from the tear edge. It was assumed that the SWCNTs were (10, 10) nanotubes, with a diameter of 1.36 nm and a wall thickness of 0.34 nm. The rope cross sections were also found to be circular.[6]

Figure 9.1 shows the SWCNT rope tensile-loading experiment, with the rope between the AFM tip and a SWCNT "paper" sample. A carbonaceous deposit can be clearly seen at the AFM tip, where it meets the rope in Fig. 9.1b. It can be seen in Fig. 9.1d that this deposit was still robust after the rope broke off. A schematic showing an overview of the tensile-loading experiment is illustrated in Fig. 9.1e, with the gray cantilever indicating where the cantilever would be if no rope were attached on the AFM tip. A parallel bimorph was added onto the stage, acting as a flexure element, to provide the force and displacement needed. The AFM probe was attached onto the bimorph and this also acted as the force sensor to determine the load.

The cross-sectional area of the SWCNT ropes can be calculated in two ways. The first is the total number of SWCNTs multiplied by the cross-sectional area of a SWCNT. However, only the perimeter SWCNTs in the rope were assumed to carry the initial load. As a result, the second and more accurate way multiplies the number of SWCNTs on the perimeter only by the cross-sectional area of a SWCNT.

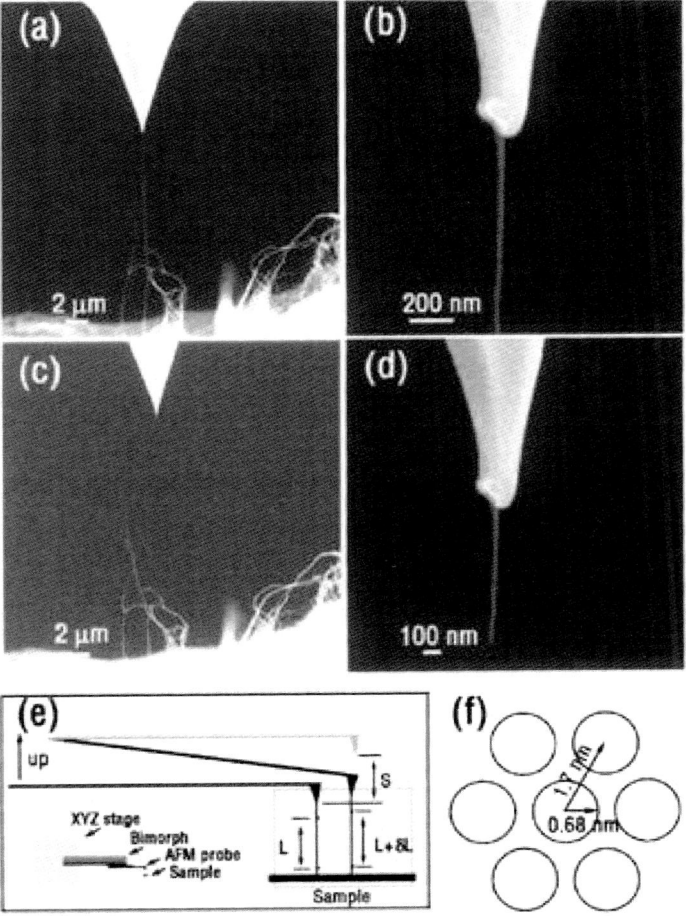

Figure 9.1 SEM images showing a SWCNT rope tensile-loading experiment:[6] (a) tensile-loaded rope; (b) close-up view of the tip; (c) rope after breaking from excessive loading; (d) close-up view of the tip after breaking; (e) schematic of the overall experiment; and (f) the ideal close-packed SWCNT rope geometry. (Reprinted with permission from Yu et al.[6])

The measured stress–strain curves from the tensile-loading experiments outlined in Fig. 9.1, for individual SWCNT ropes, are given in Fig. 9.2. Here, each rope diameter is measured at high magnification (~100,000) using a scanning electron microscope (SEM) to an accuracy of a few nanometers. The stress values have been calculated using the cross-sectional area of the perimeter SWCNT only, according to the geometric model given in Fig. 9.1f.

Using the perimeter model, from 15 SWCNT ropes it was found that the breaking strength ranged from 13 to 52 GPa (with a mean value of 30 GPa) and the Young's modulus values ranged from 320 to 1470 GPa (with a mean value of 1002 GPa). Moreover, there did not appear to be a dependence of breaking strength or Young's modulus on the rope's diameter.[6]

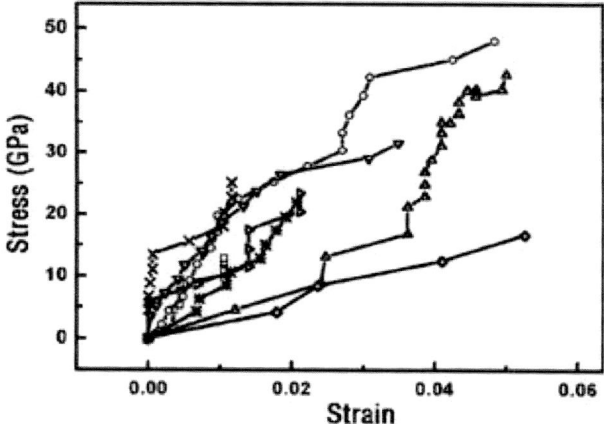

Figure 9.2 Stress–strain curves from the tensile-loading experiments of individual SWCNT ropes. (Reprinted with permission from Yu *et al.*[6])

9.4 BUCKLING TEST

Miyazawa *et al.* synthesized C_{60} nanowhiskers by means of the LLIP method[2,3] using a saturated solution of C_{60} molecules in pyridine and IPA. It was found from the electron diffraction pattern that the nanowhiskers are single crystal with a body-centered tetragonal structure.[4] The intermolecular distance of the nanowhisker was ~3% smaller than that of bulk C_{60} crystals, suggesting that polymerization occurs within the nanowhisker.

The solution containing the nanowhiskers was poured onto a microgrid that was subsequently mounted on a TEM specimen holder. A microcantilever having a gold-plated nanometer-sized silicon AFM tip was then fixed onto another specimen holder. Both holders were then inserted into the TEM so that the nanowhiskers could be deformed using the cantilever tip, as illustrated in Fig. 9.3.

The cantilever tip was attached to the tip of a 130 nm diameter C_{60} nanowhisker and then compression took place along the long axis. The force increased from 0 to 29 nN abruptly and then gradually up to 36 nN, resulting in the 7.0 μm long nanowhisker having a curvature radius of 12.7 μm corresponding to a strain of 0.005. Observations indicated that this bending was elastic.

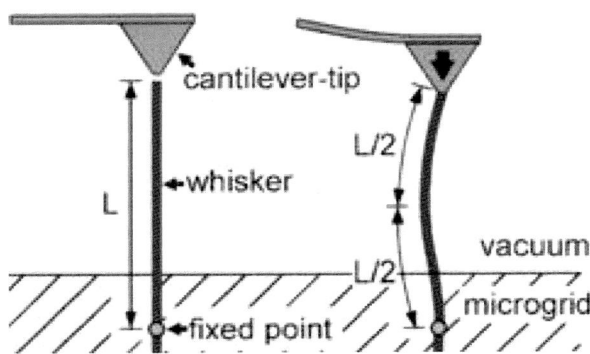

Figure 9.3 Illustration of compressive deformation of C_{60} whiskers. (Reprinted with permission from Asaka et al.,[4] © 2006, American Institute of Physics.)

Figure 9.4 shows the time-sequential TEM images of the deformation process of a 160 nm diameter, 3.3 μm long nanowhisker, leading to fracture.

Through parameter extraction, Young's modulus was estimated to be 54 and 32 GPa for 130 and 160 nm nanowhiskers, respectively.[4] When compared to the measured values of Young's modulus for bulk C_{60} crystals, with measured values ranging from 8.3 to 20 GPa, it was deduced that the increase in values for the nanowhiskers is due to the combined effects of polymerization and shape modulation.[4]

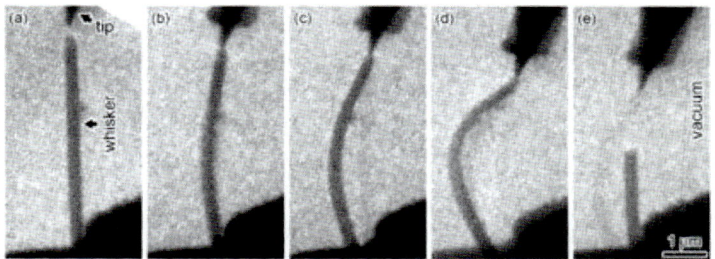

Figure 9.4 Time-sequential TEM images of the 160 nm diameter C_{60} nanowhisker leading to fracture. (Reprinted with permission from Asaka et al.,[4] © 2006, American Institute of Physics.)

9.5 MEMS BEND TESTER

A MEMS electrothermal bend tester was designed for the use under an optical microscope. The design was based around the dimensions and likely deflection range of a typical C_{60} (nano)whisker. Figure 9.5 illustrates the proposed device, containing a pair of back-to-back bend testers on a common die.

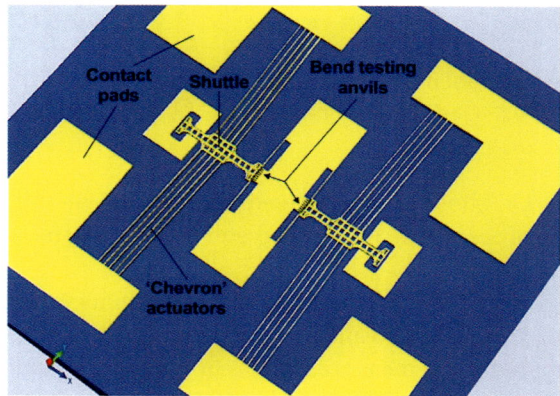

Figure 9.5 Proposed MEMS bend tester designed for C_{60} (nano)whiskers.[5]

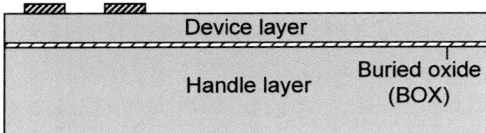
1. Photoresist application and patterning

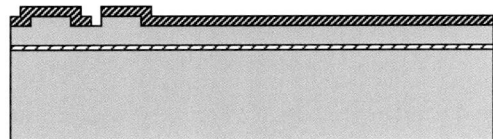
2. DRIE to form sample platform and knife-edges; photoresist application and patterning

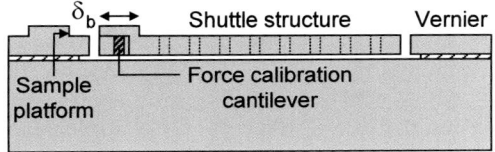
3. DRIE to BOX layer and HF etching to release compliant shuttle structure

Figure 9.6 Process flow outlining steps in the fabrication of the MEMS bend tester.[5]

Actuation is achieved by the application of current between pairs of contact pads, causing heating within "chevron" arms that support a suspended "shuttle" structure. The arms are angled in a manner to generate linear motion from their combined thermal expansion. The shuttle is attached to a set of anvils at one end, which are close to a fixed set of anvils (opposite), in order to apply bending point loads on a pre-positioned C_{60} whisker sample.

A process flow for the fabrication of the MEMS structure is shown in Fig. 9.6, relying on a two-stage deep reactive-ion etching of a bonded silicon-on-insulator wafer, to form an interdigitated set of (movable) anvil and (fixed) platform structures. Such features are important to support and raise the C_{60} (nano)whisker from the clearance zone between device and handle silicon layers, such that they can be loaded without falling into, and getting trapped within, the clearance area.

Prototypes were fabricated to test device performance and compare this with design predictions. Figure 9.7 shows optical microscope images of prototype MEMS bend testers.

The MEMS bend tester was designed for the ease of fabrication and use. To this end, features for rapid optical evaluation of force–deflection characteristics were included. Such features include a Vernier scale, with micron demarcations, and force calibration cantilevers to help in the determination of bending forces exerted on C_{60} (nano)whisker samples during testing. As the Young's modulus for silicon can be estimated to a good approximation, the force–deflection characteristics for the cantilevers can be ascertained from their dimensions. Bend tests in the absence of samples will allow the force–deflection characteristics of the bend tester to be determined as a function of applied DC power. When a C_{60} (nano)whisker sample is inserted between the anvils, the force–deflection–power characteristics will change, the difference relative to the original being the additional force applied to the sample. This approach, although simple in theory and easy to realize in a MEMS device, can be difficult to achieve in practice. The problem lies in the ability to resolve differences in force–deflection–power characteristics, under an optical microscope, from normalized values after the insertion of a sample between the loading anvils. Even if quantitative data cannot be obtained, qualitative information on the mode of fracture and the extent of deformation prior to this would be useful information.

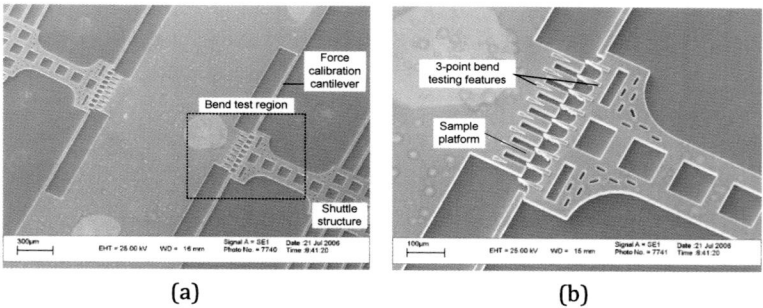

(a)   (b)

Figure 9.7 Prototype MEMS bend testers: (a) central region showing both sets of loading anvils; (b) detail over the region indicated in (a).[5]

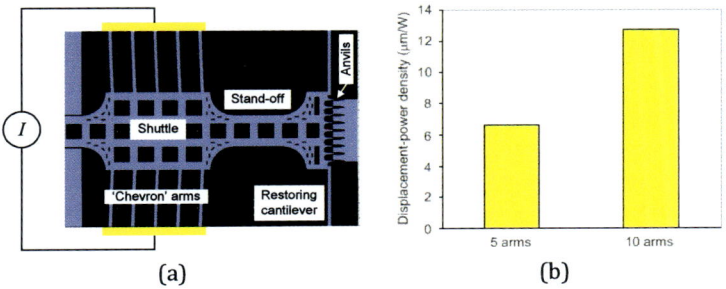

Figure 9.8 (a) Illustration showing the application of current to a tester and (b) averaged deflection–power density data for devices with 5 and 10 arms on either side of the shuttle.[5]

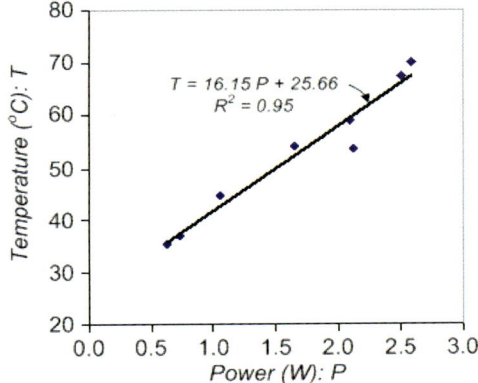

Figure 9.9 Variation in average device temperature with applied power.[5]

Bend testers with 5 and 10 supporting arms on either side of the central shuttle were fabricated to vary the force applied to different samples. The spacing between opposite anvils was varied along each tester to accommodate whiskers of different diameters and lengths. Finally, testers with anvil arrangements to perform 3- and 4-point tests were designed. The latter allows more accurate data to be obtained, whereas the former generates greater observable deflections. The displacement performance of the prototype bend testers was evaluated by passing current in a manner shown in Fig. 9.8a. The thickness of the device layer in all prototypes was 10 μm, and resultant shuttle deflections–power densities are displayed in Fig. 9.8b. For example, the graph shows that a 10-arm device produces a 12.5 μm/W deflection per Watt of the applied DC power.

During performance evaluations, an infrared camera was used to image the thermal distribution throughout the device and measure its average temperature as a function of applied power. Figure 9.9 shows the variation in average device temperature with input power on a 10-arm bend tester. At full deflection, the device reaches a maximum temperature that is marginally in excess of 70°C, corresponding to a power of approximately 2.7 W. Comparison between actual performance and design predictions revealed good agreement, showing the bend tester device to be functioning as intended. The next step involved the testing of real samples.

9.5.1 Bend Tests on C_{60} Whiskers

The placement of C_{60} nanowhiskers between the anvils of the bend testers was always going to be difficult to achieve in practice. One approach for the deposition of nanowhiskers adopted the direct application of a solvent suspension of nanowhiskers over the target areas. The disadvantage of this approach is that it is imprecise and introduces "debris" in the clearance zone between the movable shuttle and the substrate handle layer. Although the shuttle is capable of generating sufficient force to override such obstructions, extracting E from a nanowhisker sample through differentials in force–displacement–power coefficients was not possible.

A more reliable approach for the precise positioning of C_{60} nanowhiskers in regions between the anvils of the bend testers was eventually employed. The technique involved the use of a flame-stretched glass capillary tube, having a tip diameter of 5 μm. By attaching the capillary tube to a Mitutoyo x–y–z positioner, a high-resolution positioning device could be realized, as shown in Fig. 9.10. If two are used in parallel, C_{60} nanowhiskers can be moved into position with considerable ease.

In practice, the glass capillary tubes used were coated in a sputtered layer of Au to maintain a clean contact surface and promote electrostatic attraction with nanowhiskers. It is known that Au surfaces give up electrons during rubbing contact due to the triboelectric effect, generating a build-up of a positive charge. This will allow the capillary tubes to establish a positive static charge of sufficient magnitude to attract nanowhiskers, which have been shown to develop an attraction to positively charged surfaces. In some cases, this "pick-and-place" approach did not work, as van der Waal's forces between nanowhiskers and the silicon surface were too great. Such surface interactions tended to be greater with the narrowest nanowhiskers. In some cases, the nanowhiskers could be dislodged from the silicon surface using the probe tip; however, the difference in size between the probe tip and brittle nanowhisker often caused the nanowhisker to fracture. As such, the narrowest nanowhiskers could not be successfully positioned using this technique, and only samples with diameters in excess of 0.5 µm could be placed between the anvils of the bend testers.

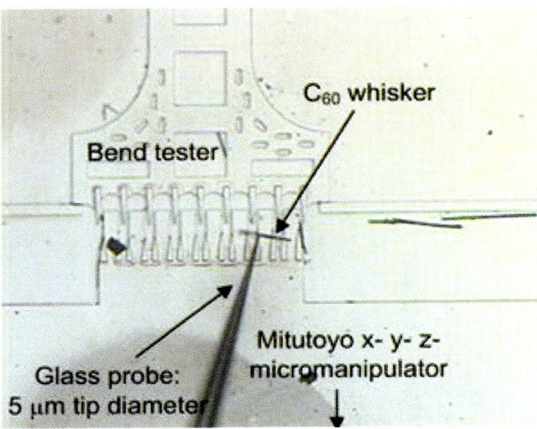

Figure 9.10 Placement of a C_{60} whisker between the anvils of a bend tester, using a micropositioner attached to a capillary tube having a 5 µm tip diameter.[5]

Optical microscope images of samples in 3- and 4-point bend testers are shown in Figs. 9.11 and 9.12, respectively. A fractured C_{60} nanowhisker is seen in Fig. 9.11a, following a prior successful 3-point bend test. Failure appears to be entirely brittle, as no signs of plastic deformation are visible. Figure 9.11b shows the same tester during the application of DC power. The larger C_{60} whisker, to the right of the C_{60} nanowhisker, is now experiencing loading between the three contacting anvils. Repeated loading and unloading revealed elastic deformation characteristics. Unfortunately, in the process of this basic "fatigue" test, the sample worked its way above the anvils and could not be tested to destruction.

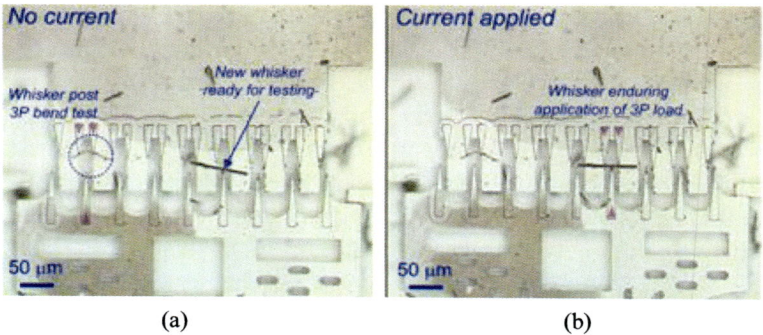

Figure 9.11 3-point bend test of C_{60} whiskers: (a) before and (b) during tester actuation.[5]

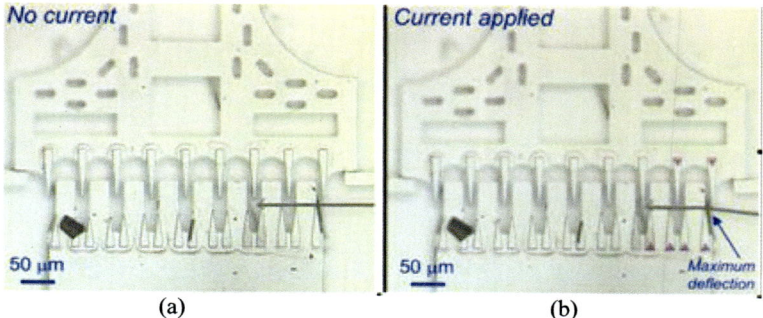

Figure 9.12 4-point bend test of a C_{60} whisker: (a) before and (b) during tester actuation.[5]

Closer inspection of the image in Fig. 9.12b reveals that the loading anvil at the far right of the tester is marginally longer than the other three anvils in contact with the whisker. So despite the fact that the whisker is clearly under load from all four anvils, the greatest deformation (and, hence stress) is occurring in the region above the third from the left anvil. The unusual situation conveniently produces what is in effect a cantilever bend test, with the "fixed" support provided by anvils 1–3 (left to right) and the point load applied by anvil 4. Figure 9.13 shows a simplified schematic representation of the cantilever bend test between the third and fourth anvils.

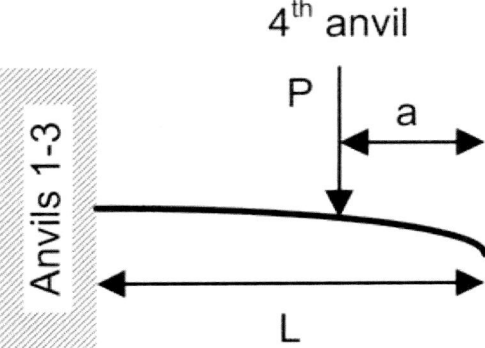

Figure 9.13 Cantilever bending due to a transverse point load, representing a possible scenario for loading between anvils 3 and 4 of the bend tester in Fig. 9.12.[5]

The governing equation that relates the deflection of the cantilever's tip is given by[7]

$$\delta = \frac{P}{6EI}\left(2L^3 - 3L^2 a + a^3\right) \qquad (9.1)$$

where δ is the tip deflection, I is the second moment of area of the cantilever, and P is the applied load. The image in Fig. 9.12b was captured at the instant immediately prior to sample fracture, which occurred at the fourth anvil. The deflection observed is the maximum for the sample prior to fracture. The distance between the third and fourth anvils is approximately 10 µm, the maximum deflection noted at the point of load application was measured at just under 1.5 μm, and the whisker's diameter was determined to be approximately 4 μm. Figure 9.14 shows the fractured whisker following successful bend testing. The fracture point indicates brittle failure.

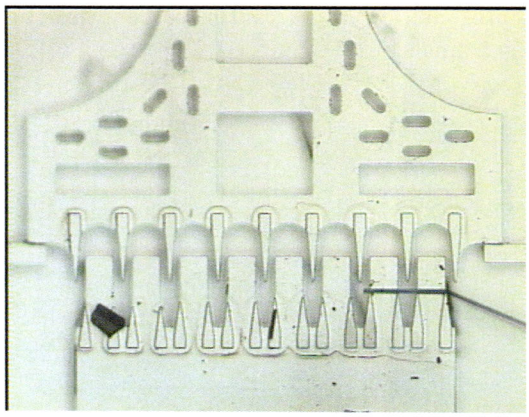

Figure 9.14 C_{60} whisker showing signs of brittle fracture following a bend test.[5]

9.5.2 Experimental Results

The resolution of the optical microscope and deflection measurement system used was not sufficient to allow representative load force data to be inferred from the applied DC power, as initially envisaged. An approximation for the Young's modulus of the C_{60} whisker in Fig. 9.12 is given, assuming a maximum resolution for the optical system of 1 μm. If there is no change observed in the force–deflection–power characteristics of the bend tester following sample insertion, the additional load force applied to the sample corresponds to that required to move the unloaded tester through a distance of ~1 μm. Based on the dimensions of the force calibration cantilevers and an assumed E for silicon of 160 GPa, this corresponds to a force of ~2.5 μN; in other words, the bend tester has a force–deflection characteristic of 2.5 μN/μm. Assuming this force to be of the same order of magnitude to the maximum load applied to the whisker in the bend test, a simple calculation using Eq. 9.1 yields an approximation of its stiffness and, hence, E. By assuming the whisker's shape to follow that of a solid rod of diameter D, the second moment of area, I, is calculated using the following standard expression:

$$I = \frac{\pi D^4}{64} \tag{9.2}$$

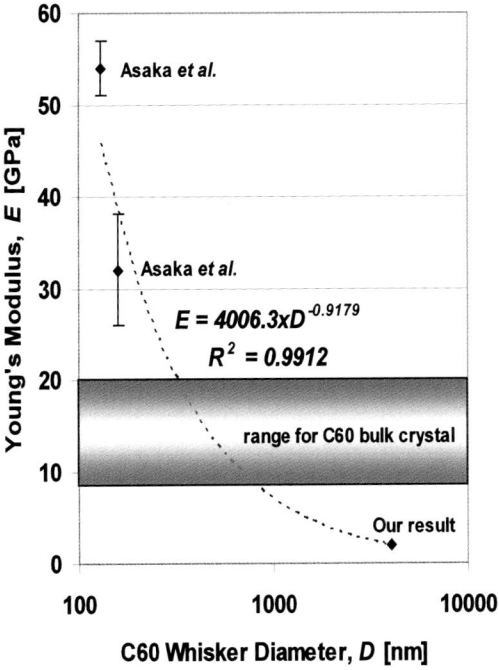

Figure 9.15 Characterization of Young's modulus with C_{60} whisker diameter.[5]

Although the whisker's tip deflection is cropped from the image in Fig. 9.12b, its value was measured. With remaining measurements on the loaded sample inserted into Eq. 9.1, the parameter extraction method suggests a value of $E \sim 2$ GPa for the 4 μm diameter C_{60} whisker. This value is at least a factor of 10 lower than that reported by Asaka et al.,[4] with their axially loaded C_{60} nanowhisker having a diameter of 160 nm. Our C_{60} whisker has a diameter of 4 μm, which is at least 25 times thicker than those reported by Asaka et al.[4] This makes it an interesting sample for comparison. From the limited experimental data available (two data points from Asaka et al. and one from our work), Young's modulus can be crudely modeled, as shown in Fig. 9.15.

Ideally, more data points from controlled samples are needed in order to determine a more definitive model, but the inverse proportionality of Young's modulus to C_{60} whiskers diameter is evident. This appears to contradict any notion of a fixed value of Young's modulus for C_{60} whiskers.

9.5.3 Discussion

The lower stiffness of the whisker characterized in this study is likely to be a consequence of the presence of grain boundaries due to its polycrystalline structure. Of course, another possibility is that the technique employed leads to an underestimation of the stiffness of the whisker. One possible source of error is the measurement of maximum whisker deflection prior to failure. Although previous deflections can equally be used to determine E, the final value was used as this facilitated measurement through the optical microscope. Other, more minor, errors may come from the approximation leading to the scenario represented by Fig. 9.13 and subsequent use of Eq. 9.1.

Despite the limitations of the mechanical characterization carried out in this study, the results are informative on qualitative grounds as they provide evidence for brittle fracture in C_{60} nanowhiskers as well as larger diameter whiskers.

9.6 CONCLUSIONS

Little has been published on the mechanical bend testing of fullerene (nano)whiskers due to the inherent difficulties in physically mounting such small test samples. The tensile loading method for CNTs, the bucking test method for C_{60} nanowhiskers, and the MEMS bend tester for C_{60} (nano)whiskers have been introduced. All three methods are far from ideal and their findings are subject to further investigation.

With the results from the bucking test and MEMS testing of C_{60} (nano)whiskers, combining the three data points, there is the suggestion of an inverse proportionality relationship between the Young's modulus and the diameter of a C_{60} whisker.[5]

Acknowledgments

The authors are very grateful to Dr. Kun'ichi Miyazawa for providing the initial inspiration for our work on C_{60} whiskers and ongoing discussions, and also Dr. Adam Wojcik at University College London. We also like to thank the UK's Engineering and Physical Research Council for funding this research (under grant GR/S97019/01).

References

1. H. W. Kroto, J. R. Heath, S. C. O'Brien, R. F. Curl, and R. E. Smalley, *Nature*, 318, 162 (1985).
2. K. Miyazawa, Y. Kuwasaki, A. Obayashi, and M. Kuwabara, *J. Mater. Res.*, **17**, 83 (2002).
3. K. Miyazawa, Y. Kuwasaki, K. Hamamoto, S. Nagata, A. Obayashi, and M. Kuwabara, *Surf. Interface Anal.*, **35**, 117 (2003).
4. K. Asaka, R. Kato, K. Miyazawa, and T. Kizuka, *Appl. Phys. Lett.*, **89**, 071912 (2006).
5. M. P. Larsson and S. Lucyszyn, *J. Phys.: Conf. Ser.*, **159**, 012006 (2009).
6. M. F. Yu, B. S. Files, S. Arepalli, and R. S. Ruoff, *Phys. Rev. Lett.*, **84**, 5552 (2000).
7. W. C. Young and R. G. Budynas, *Roark's Formulas for Stress and Strain* (McGraw-Hill International Edition, New York, 2002).

Chapter 10

MAGNETIC ALIGNMENT OF FULLERENE NANOWHISKERS

Guangzhe Piao,[a,b] Fumiko Kimura,[c] and Tsunehisa Kimura[c]

[a] *Key Laboratory of Rubber-Plastics (QUST), Ministry of Education, Qingdao University of Science and Technology, 53 Zhengzhou Road, Qingdao 266042, People's Republic of China*
[b] *School of Polymer Science and Engineering, Qingdao University of Science and Technology, 53 Zhengzhou Road, Qingdao 266042, People's Republic of China*
[c] *Division of Forest and Biomaterials Science, Graduate School of Agriculture, Kyoto University, Kitashirakawa, Sakyo-ku, Kyoto 606-8502, Japan*
piao@qust.edu.cn; tkimura@kais.kyoto-u.ac.jp

The magnetic alignment of the C_{60} fullerene nanowhiskers suspended in poly(vinyl alcohol) solution under magnetic fields of 10–12 T at room temperature was investigated. The result is very interesting and showed that C_{60} nanowhisker was oriented with the long axis parallel to the field while the one with tubular structure was aligned perpendicular to the field.

Fullerene Nanowhiskers
Edited by Kun'ichi Miyazawa
Copyright © 2012 Pan Stanford Publishing Pte. Ltd.
www.panstanford.com

10.1 INTRODUCTION

The discovery of C_{60} fullerenes[1] and the method used to manufacture bulk fullerenes in macroscopic quantities[2] have triggered a great deal of interest among scientists from both fundamental and practical point of views because of their high symmetry, novel electropool π-conjugated system, and unique chemical and physical properties. Fullerene C_{60} molecules usually crystallize in films and plates. Optical limiting devices,[3-6] organic solar cells,[7-11] superconductors,[12-14] organic thin film transistors,[15-17] electrical conductivity, as well as hardness and deformation of the crystalline C_{60} films and plates[18-21] have been reported in recent decades. It has been known that various structures and shapes of fullerene C_{60} crystals are produced from their liquid solutions, depending on the organic solvent used, growth temperature, and so on.[22-27] Stimulated by the discovery of the carbon nanotube,[28] intensive researches on the synthesis and investigation of one-dimensional (1D) fullerene nanowires such as fullerene nanowhiskers (FNWs, without tubular structure) and fullerene nanotubes (FNTs, tubular nanowires) are being preformed.[29-44] Recently, Miyazawa et al.[37-40] developed a simple liquid–liquid interfacial precipitation method (LLIP) to produce 1D single crystalline fullerene nanowhiskers mainly composed of C_{60} molecules with the length larger than several millimeters and the diameters ranging from submicron to 1 μm. The nanowhiskers with high aspect ratio might be of potential importance, such as single-walled carbon nanotubes, for nanometer-scale devices, which show various electronic properties from semiconducting to metallic behaviors depending on their chirality and diameter.

In many aspects, control of alignment is of great importance. For example, if the long axes of π-conjugated molecules are uniaxially aligned in a film, the resultant emission and absorption will be highly anisotropic because the transition dipole moments lie parallel to the long axis.[45-51] Therefore, the alignment of 1D fullerene C_{60} nanowhiskers, in a particular direction, imparts specific functionality to the film. On the other hand, fibers with diamagnetic anisotropy align under static magnetic fields. Carbon fibers,[52,53] carbon nanotubes,[54-57] polyethylene fibers,[58] cellulose fibers,[59,60] and so forth undergo magnetic alignment.[61] The alignment occurs so that the axis of the largest diamagnetic susceptibility lies parallel to the applied field. The alignment manner depends on the anisotropic nature of these fibers. For example, fibers with positive diamagnetic anisotropy ($\chi_a = \chi^\parallel - \chi^\perp > 0$) undergo uniaxial alignment, i.e., they align with their fiber axes parallel to the applied field, while those with negative diamagnetic anisotropy ($\chi_a < 0$) undergo planar alignment. Here $\chi^\parallel$ and $\chi^\perp$ are the diamagnetic susceptibilities in the directions parallel and perpendicular to the fiber axis, respectively. Generally, carbon fibers and carbon nanotubes belong to the former, while polyethylene and cellulose fibers belong to the latter.

In the present chapter, the magnetic alignment of the fullerene C_{60} nanowhiskers in poly(vinyl alcohol) solution is reported. The results showed that FNW was oriented with the long axis parallel to an external magnetic field while FNT was aligned perpendicular to the field.

10.2 EXPERIMENTAL

C_{60} nanowhiskers were prepared by using the LLIP method.[37-40] As purchased, C_{60} (>99%, MTR Ltd., Cleveland, OH, USA) was dissolved in pyridine sonically. A saturated pyridine solution of C_{60} was irradiated with 400–500 nm light for 2 h. Then 18 mL of 2-propanol was added to each of the two glass bottles containing 2 mL of the saturated pyridine solution of C_{60}. To obtain suitable diffusion at the interface, ultrasonic dispersion was performed after addition of 2-propanol for 1 min. The solutions of C_{60} were kept at −22 and 8.8°C overnight, respectively. The C_{60} FNWs were grown at −22°C and sank to the bottom of the bottle. Also, C_{60} FNTs were obtained at 8.8°C. The formed fine fibrous precipitates of C_{60} in the glass bottle were pipetted into a glass bottle containing 2 mL of 2-propanol and sonicated for preparing short nanowhiskers. The above short nanofibers in the glass bottle were pipetted into a vessel containing poly(vinyl alcohol) solution [5 wt % poly(vinyl alcohol) in water] whose bottom was covered with a cover glass. The vessel containing the suspension was placed at the center of a JASTEC cryogen-free superconducting magnet generating a horizontal field of 10–12 T, and the solvent was evaporated at room temperature to prepare polymer nanocomposites containing magnetically aligned C_{60} nanowhiskers. Morphological observations were performed using a polarizing optical microscope (POM; Olympus) connected to a digital camera, a scanning electron microscope (SEM; JEOL JSM-6700F), and a transmission electron microscope (TEM; JEOL, JEM-2000EX). The specimens in the glass bottles were dropped on a slide glass or mounted onto a microgrid after ultrasonically irradiating for 1 min.

10.3 RESULTS AND DISCUSSION

10.3.1 Morphologies

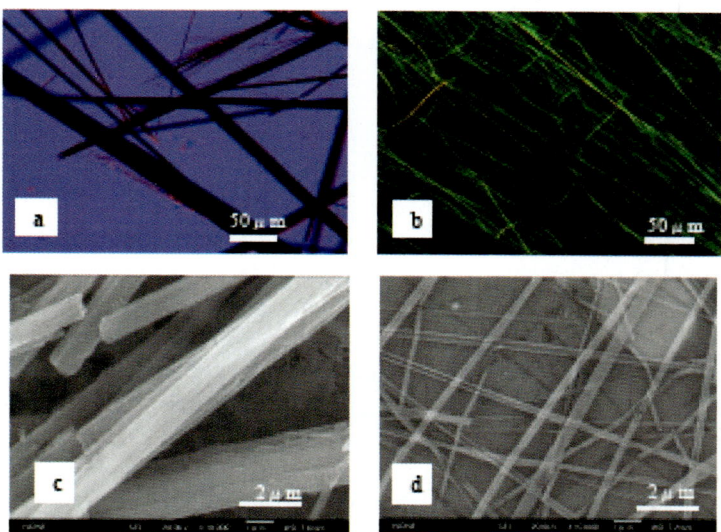

Figure 10.1 POM and SEM images of (a, c) C_{60} FNWs and (b, d) C_{60} FNTs, respectively.

Figure 10.1 shows POM and SEM photographs of the prepared FNWs and FNTs. As shown in Fig. 10.1, the C_{60} fibers were submicron in diameter and reached several millimeters in length, i.e., FNWs. As can be seen from Fig. 10.1a, c, the FNWs form bundles consisting of several tens of fine nanowires with several to tens micrometers in diameter and several millimeters in length. On the other hand, the C_{60} FNTs formed single tubular fibers with about 200 nm of outer diameters and several millimeters in length (Fig. 10.1b, d). Figure 10.2 shows TEM images of the C_{60} FNWs that were taken from the glass bottle 1 min after the pulverization. By the pulverization, open edges are observed to confirm the tubular structure of the specimen. In Fig. 10.2, the outer diameter of FNT was about 160 nm and that of FNW was about 90 nm.

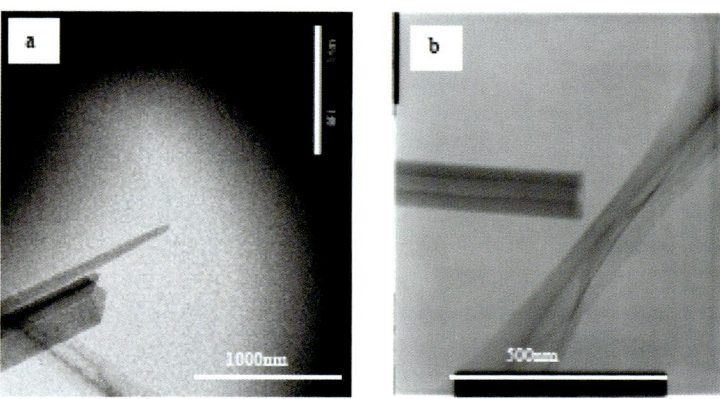

Figure 10.2 TEM images of (a) C_{60} FNWs and (b) C_{60} FNTs, respectively.

10.3.2 Magnetic Alignment of C_{60} FNWs

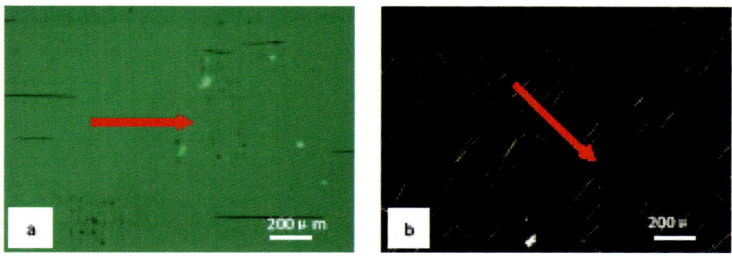

Figure 10.3 POM images of magnetic alignment of (a) C_{60} FNWs and (b) C_{60} FNTs dispersed in poly(vinyl alcohol) solution under 10–12 T magnetic field, respectively. The field direction is shown with an arrow.

The suspension of C_{60} FNWs was placed in magnetic fields, and the solvent was vaporized. The alignment was observed by POM. As shown in Fig. 10.3a, C_{60} FNWs were oriented with the fibril axis parallel to the magnetic field. On the other hand, as shown in Fig. 10.3b, C_{60} FNTs were aligned with the tube axis perpendicular to the field. The magnetic orientation of the C_{60} FNWs is explained by the anisotropic magnetic susceptibility. FNWs are magnetically symmetric along the fibril axis and possess susceptibilities parallel ($\chi^{\parallel}$) and perpendicular ($\chi^{\perp}$) to it. The magnetic energy of C_{60} FNWs with volume V placed in a magnetic field B is expressed by $E(\theta, B) = -(VB^2/2)[\chi^{\perp} + (\chi^{\parallel} - \chi^{\perp})\cos^2\theta]$, where θ is the angle between the fibril axis and the field H. The magnetic alignment occurs so that the energy $E(\theta, B)$ becomes minimum. Experimentally, C_{60} FNWs were aligned with the fibril axis parallel to the field ($\theta = 0°$) and the C_{60} FNTs were oriented with the tube axis perpendicular to the field ($\theta = 90°$). Because they are considered to be diamagnetic at room temperature, this requires a condition of $\chi^{\perp} < \chi^{\parallel} < 0$ for the C_{60} FNWs and a condition of $\chi^{\perp} > \chi^{\parallel}$ for the FNTs. At the present time, the origin of the opposite anisotropic susceptibilities between the C_{60} FNWs and the FNTs is not fully clarified. Both C_{60} FNWs and FNTs have a hexagonal crystal structure but turn to a face-centered cubic structure when dried in air. But the structure of C_{60} FNWs in solution was different from that reported for the FNTs. Probably the anisotropic susceptibility of C_{60} FNWs was very sensitive to the solvated crystal structure.

10.4 SUMMARY

A facile method for the alignment of the C_{60} FNWs suspended in poly(vinyl alcohol) solution under magnetic fields at room temperature has been demonstrated. The achieved alignment in a suspension was fixed by removing the suspending liquid by evaporation. The different magnetic alignment behaviors between FNWs and FNTs can be attributed to their different solvated crystal structures. Clearly, many open questions such as size-dependent magnetic anisotropy, the outer oxide layers of C_{60} FNWs, and the possible influence of solvent used for preparing whisker suspensions should be answered experimentally and theoretically in detail to fully understand the magnetic properties of the C_{60} FNWs.

Acknowledgments

The authors are grateful to Dr. Kun'ichi Miyazawa and Dr. Jun-ichi Minato for valuable discussion on this work. This work was partially supported by the Grant-in-Aid for Scientific Research on Priority Area "Innovative utilization of strong magnetic fields" (Area 767, No. 15085207) from MEXT of Japan, the Natural Science Foundation of China (Nos. 50773033 and 50872060), Science Foundation of Shandong Province (Nos. Y200701 and Q2008F07), and Doctoral Foundation of QUST (2008).

References

1. H. W. Kroto, J. R. Heath, S. C. O'Brien, R. F. Curl, and R. E. Smalley, *Nature*, **318**, 162 (1985).
2. W. Krätschmer, L. D. Lamb, K. Fostiropoulos, and D. R. Huffman, *Nature*, **347**, 354 (1990).
3. L. W. Tutt and A. Kost, *Nature*, **356**, 225 (1992).
4. Y. P. Sun and J. E. Riggs, *Int. Rev. Phys. Chem.*, **18**, 43 (1999).
5. G. Brusatin and R. Signorini, *J. Mater. Chem.*, **12**, 1964 (2002).
6. J. F. Nierengarten, N. Armaroli, G. Accorsi, Y. Rio, and J. F. Eckert, *Chem. Eur. J.*, **9**, 37 (2003).
7. J. F. Nierengarten, *New J. Chem.*, **28**, 1177 (2004).
8. Y. Kim, S. Cook, S. M. Tuladhar, S. A. Choulis, J. Nelson, J. R. Durrant, D. D. C. Bradley, M. Giles, I. McCulloch, C. S. Ha, and M. Ree, *Nat. Mater.*, **5**, 197 (2006).
9. C. J. Brabec, N. S. Sariciftci, and J. C. Hummelen, *Adv. Funct. Mater.*, **11**, 15 (2001).
10. J. F. Nierengarten, *Sol. Energy Mater. Sol. Cells*, **83**, 187 (2004).
11. J. L. Segura, N. Martin, and D. M. Guldi, *Chem. Soc. Rev.*, **34**, 31 (2005).
12. M. Capone, M. Fabrizio, C. Castellani, and E. Tosatti, *Science*, **296**, 2364 (2002).

13. Z. Zhang, C. C. Chen, S. P. Kelty, H. J. Dai, and C. M. Lieber, *Nature*, **353**, 333 (1991).
14. Z. Iqbal, R. H. Baughman, B. L. Ramakrishna, S. Khare, N. S. Murthy, H. J. Bornemann, and D. E. Morris, *Science*, **254**, 826 (1991).
15. T. Nagano, H. Sugiyama, E. Kuwahara, R. Watanabe, H. Kusai, Y. Kashino, and Y. Kubozono, *Appl. Phys. Lett.*, **87**, 023501 (2005).
16. C. D. Dimitrakopoulos and P. R. L. Malenfant, *Adv. Mater.*, **14**, 99 (2002).
17. R. C. Haddon, A. S. Perel, R. C. Morris, T. T. M. Palstra, A. F. Hebard, and R. M. Fleming, *Appl. Phys. Lett.*, **67**, 121 (1995).
18. R. M. Fleming, A. R. Kortan, B. Hessen, T. Siegrist, F. A. Thiel, P. Marsh, R. C. Haddon, R. Tycko, G. Dabbagh, M. L. Kaplan, and A. M. Mujsce, *Phys. Rev. B*, **44**, 888 (1991).
19. Y. Yoshida, *Jpn. J. Appl. Phys.*, **31**, L505 (1992).
20. H. Moriyama, H. Kobayashi, A. Kobayashi, and T. Watanabe, *Chem. Phys. Lett.*, **238**, 116 (1995).
21. R. Ceolin, J. L. Tamarit, D. O. Lopez, M. Barrio, V. Agafonov, H. Allouchi, F. Moussa, and H. Szwarc, *Chem. Phys. Lett.*, **314**, 21 (1999).
22. S. Ogawa, H. Furusawa, T. Watanabe, and H. Yamamoto, *J. Phys. Chem. Solids*, **61**, 1047 (2000).
23. J. E. Fishcher, P. A. Heiney, and A. B. Smith, *Acc. Chem. Res.*, **25**, 112 (1992).
24. F. Diederich and M. Gomez-Lopez, *Chem. Soc. Rev.*, **28**, 263 (1999).
25. T. D. Ros and M. Prato, *Chem. Commun.*, 663 (1999).
26. A. F. Hebard, M. J. Rosseinsky, R. C. Haddon, D. W. Murphy, S. H. Glarum, T. T. M. Palstra, A. P. Ramirez, and A. R. Kortan, *Nature*, **350**, 600 (1991).
27. Y. Kim, S. Cook, S. M. Tuladhar, S. A. Choulis, J. Nelson, J. R. Durrant, D. D. C. Bradley, M. Giles, I. Mcculloch, C.-S. Ha, and M. Ree, *Nat. Mater.*, **5**, 197 (2006).
28. S. Iijima, *Nature*, **354**, 56 (1991).
29. M. F. Meidine, P. B. Hitchcock, H. W. Kroto, R. Taylor, and D. R. M. Walton, *J. Chem. Soc., Chem. Commun.*, 1534 (1992).
30. Y. Yosida, *Jpn. J. Appl. Phys.*, **31**, L505 (1992).

31. L. Balch, J. W. Lee, B. C. Noll, and M. M. Olmstead, *J. Chem. Soc., Chem. Commun.*, 56 (1993).
32. V. Talyzin, *J. Phys. Chem. B*, **101**, 9679 (1997).
33. R. Swietlik, P. Byszewski, and E. Kowalska, *Chem. Phys. Lett.*, **254**, 73 (1996).
34. S. Pekker, G. Faigel, K. Fodor-Csorba, L. Granasy, E. Jakab, and M. Tegze, *Solid State Commun.*, **83**, 423 (1992).
35. R. Ceolin, V. Agafonov, B. Bachet, A. Gonthier-Vassal, H. Szwarc, S. Toscani, G. Keller, C. Fabre, and A. Rassat, *Chem. Phys. Lett.*, **244**, 100 (1995).
36. S. Toscani, H. Allouchi, J. L1. Tamarit, D. O. Lopez, M. Barrio, V. Agafonov, A. Rassat, H. Szwarc, and R. Ceolin, *Chem. Phys. Lett.*, **330**, 491 (2000).
37. K. Miyazawa, A. Obayashi, and M. Kuwabara, *J. Am. Ceram. Soc.*, **84**, 3037 (2001).
38. K. Miyazawa, Y. Kuwasaki, A. Obayashi, and M. Kuwabara, *J. Mater. Res.*, **17**, 83 (2002).
39. J. Minato and K. Miyazawa, *Carbon*, **43**, 2837 (2005).
40. J. Minato, K. Miyazawa, and T. Suga, *Sci. Technol. Adv. Mater.*, **6**, 272 (2005).
41. L. Wang, B. Liu, D. Liu, M. Yao, Y. Hou, S. Yu, T. Cui, D. Li, G. Zau, A. Iwasiewicz, and B. Sundqvist, *Adv. Mater.*, **18**, 1883 (2006).
42. L. Wang, B. Liu, S. Yu, M. Yao, D. Liu, Y. Hou, T. Cui, G. Zau, B. Sundqvist, H. You, D. Zhang, and D. Ma, *Chem. Mater.*, **18**, 4190 (2006).
43. Y. Jin, R. J. Curry, J. Sloan, R. A. Hatton, L. C. Chong, N. Blanchard, V. Stolojan, H. W. Kroto, and S. R. P. Silva, *J. Mater. Chem.*, **16**, 3715 (2006).
44. S. Malik, N. Fujita, P. Mukhopadhyay, Y. Goto, K. Kaneko, T. Ikeda, and S. Shinkai, *J. Mater. Chem.*, **17**, 2454 (2007).
45. M. Grell, W. Knoll, D. Lupo, A. Meisel, T. Miteva, D. Neher, H.-G. Nothofer, U. Scherf, and A. Yasuda, *Adv. Mater.*, **11**, 671(1999).
46. M. Grell and D. D. C. Bradley, *Adv. Mater.*, **11**, 895 (1999).
47. M. Jandke, P. Strohriegl, J. Gmeiner, W. Brűtting, and M. Schwoerer, *Adv. Mater.*, **11**, 1518 (1999).

48. M. Era, T. Tsutsui, and S. Saito, *Appl. Phys. Lett.*, **67**, 2436 (1995).
49. T. W. Hagler, K. Pakbaz, K. F. Voss, and A. J. Heeger, *Phys. Rev. B*, **44**, 8652 (1991).
50. Y. Yoshida, N. Tanigaki, K. Yase, and S. Hotta, *Adv. Mater.*, **12**, 1587 (2000).
51. Y. Kim, N. Minami, and S. Kazaoui, *Appl. Phys. Lett.*, **86**, 073103 (2005).
52. V. J. Timbrell, *Appl. Phys.*, **43**, 4839 (1972).
53. Y. Schmitt, C. Paulick, F. X. Royer, and J. G. Gasser, *J. Non-Cryst. Solids*, **139**, 205 (1996).
54. M. Fujiwara, E. Oki, M. Hamada, Y. Tanimoto, I. Mukouda, and Y. Shimomura, *J. Phys. Chem. A*, **105**, 4383 (2001).
55. T. Kimura, H. Ago, M. Tobita, S. Ohshima, M. Kyotani, and M. Yumura, *Adv. Mater.*, **14**, 1380 (2002).
56. M. J. Casavant, D. A. Walters, J. J. Schmidt, and R. E. Smalley, *J. Appl. Phys.*, **93**, 2153 (2003).
57. G. Piao, F. Kimura, T. Takahashi, Y. Moritani, H. Awano, S. Nimori, K. Tsuda, K. Yonetake, and T. Kimura, *Polym. J.*, **39**, 589 (2007).
58. T. Kimura, M. Yamato, W. Koshimizu, M. Koike, and T. Kawai, *Langmuir*, **16**, 858 (2000).
59. J. Sugiyama, H. Chanzy, and G. Maret, *Macromolecules*, **25**, 4232 (1992).
60. T. Kimura, M. Yoshino, T. Yamane, M. Yamato, and M. Tobita, *Langmuir*, **20**, 5669 (2004).
61. T. Kimura, *Polym. J.*, **35**, 823 (2003).

Chapter 11

OPTICAL PROPERTIES OF FULLERENE NANOWHISKERS

Kiyoto Matsuishi

Institute of Materials Science, University of Tsukuba, Tsukuba,
Ibaraki 305-8573, Japan
kiyoto@bk.tsukuba.ac.jp

Optical properties of nanoscaled phases of solid C_{60} are discussed. First, electronic and optical properties of C_{60} molecules and bulk crystals are briefly reviewed. Then, optical properties of C_{60} nanocrystals, nanowhiskers, and nanotubes are presented. Finally, photo-induced structural transformations of C_{60} nanowhiskers and nanotubes to polymeric and graphitic phases are described. It is pointed out that the electronic structures, and therefore the optical properties, of C_{60} are influenced strongly by its aggregated nanoscaled forms. The condensation of C_{60} molecules into nanoscaled forms play an intriguing role in the vibronic coupling of C_{60}.

11.1 INTRODUCTION

One of the great interests in physical properties of fullerenes is to

---

Fullerene Nanowhiskers
Edited by Kun'ichi Miyazawa
Copyright © 2012 Pan Stanford Publishing Pte. Ltd.
www.panstanford.com

clarify how the electronic structure can be tailored by condensing them into various solid forms. Although physical properties of C_{60} in its molecular and bulk solid states have been investigated extensively, the properties of intermediate state between bulk and molecular states, i.e., mesoscopic or aggregated nanoscaled state of solid C_{60}, have not been understood well. Martin *et al.* observed the icosahedral $(C_{60})_n$ clusters ($10 < n < 100$) in the photoionization time-of-flight mass spectra in vacuum.[1] Aggregation of C_{60} molecules to form $(C_{60})_n$ clusters ($1 < n < 10$) was also observed in solution.[2,3] Some attempts were made to incorporate C_{60} into the pores of zeolites[4–6] and porous glasses.[7] However, those C_{60} clusters are too small to understand the properties of the mesoscopic state of solid C_{60}. It has been necessary to make stable and well-dispersed nanometer-sized C_{60} crystals to investigate the electronic properties of the mesoscopic phase of solid C_{60}.

Well-dispersed $(C_{60})_n$ nanocrystals ($1000 < n < 10{,}000$, hereafter referred to as C_{60}NCs) were prepared in an optically transparent matrix by using an inert gas evaporation method,[8] and the size effect on the optical properties of C_{60} crystals were reported.[9] Recently, stable and well-dispersed one-dimensional nanoscaled C_{60} nanowhiskers (hereafter C_{60}NWs) and nanotubes (hereafter C_{60}NTs) have been fabricated by using a liquid–liquid interfacial precipitation method.[10,11] Both C_{60}NWs and C_{60}NTs have drawn attention as quasi-one-dimensional nanomaterials due to their potential application to nanotechnology.

In this chapter, the optical properties of such a nanoscaled phase of solid C_{60} are discussed. It will be pointed out that the electronic structures, and therefore the optical properties, of C_{60} are influenced strongly by its aggregated nanoscaled forms. The condensation of C_{60} molecules into nanoscaled forms plays an intriguing role in the vibronic coupling of C_{60}. First, in this chapter, electronic and optical properties of C_{60} molecules and bulk crystals are briefly reviewed. Then, optical properties of C_{60}NCs, C_{60}NWs, and C_{60}NTs are presented. Finally, photo-induced structural transformations of C_{60}NWs and C_{60}NTs to polymeric and graphitic phases are described.

11.2 OPTICAL PROPERTIES OF ISOLATED AND SOLID C_{60}

Various optical investigations have been reported for the electronic states of the isolated and solid C_{60}. Many researchers have especially attempted to unravel the photoluminescence (PL) of solid C_{60}. Whereas consensus seems to exist with regard to the experimental PL spectrum of an isolated C_{60} molecule, the PL spectra of solid C_{60} have been controversial. The PL spectra of isolated C_{60} molecules such as C_{60} in solvent[12–16] and in rare gas matrices[17–20] have been reported. The PL spectra of isolated C_{60} molecules exhibit very sharp vibronic structures at 1.2 K and a relatively very weak electronic origin (0–0 transition). The pioneering quantitative interpretations for the vibronic coupling of C_{60} have been made by Negri et al.[21,22] They have shown that a large number of the vibronic structures can be attributed to the Hertzberg–Teller active modes and combinations of Hertzberg–Teller and Jahn–Teller active modes. In the Hertzberg–Teller mechanism, the parity-forbidden electronic transition is relaxed by the symmetry of phonon while the Jahn–Teller mechanism involves the relaxation of the symmetry selection rules of the optically forbidden transition due to the distortion of the lattice.

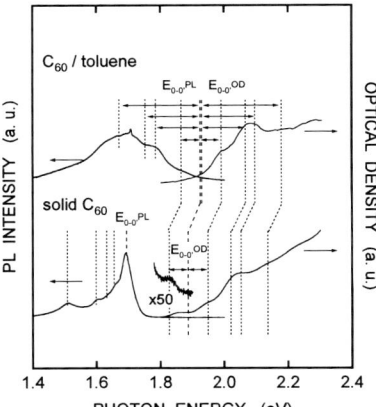

Figure 11.1 Optical absorption and PL spectra near the absorption edge (~1.9 eV) of solid C_{60} (bottom) and C_{60} in toluene (top). The spectra of the solid C_{60} (film) were taken at 80 K. The spectra of C_{60} in toluene were taken at room temperature.

Figure 11.1 shows fine structures observed near the absorption edge (~1.9 eV) in the optical absorption and PL spectra of solid C_{60} (film) and C_{60} molecules in toluene. The fine structures in toluene have a mirror symmetry at the center of 1.92 eV, which indicates that the HOMO–LUMO (highest occupied molecular orbital–lowest unoccupied molecular orbital) transition energy is about 1.92 eV. A schematic energy level diagram of an isolated C_{60} molecule is shown in Fig. 11.2.

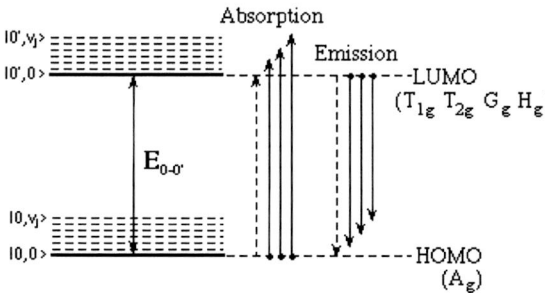

Figure 11.2 Schematic energy level diagram of an isolated C_{60} molecule. The horizontal solid and dashed lines represent electronic and vibronic states, respectively.

The isolated C_{60} molecule with the I_h symmetry exhibits 46 distinct vibration modes out of 174 internal degrees of freedom. The irreducible representations are given as

$$\Gamma_{I_h} = 2a_g + a_u + 3t_{1g} + 4t_{2g} + 4t_{1u} + 5t_{2u} + 6g_g + 6g_u + 8h_g + 7h_u.$$

The $2a_g + 8h_g$ modes are Raman active, and the four t_{1u} modes are infrared active. According to the many electron molecular approach,[23] the ground electronic state (the HOMO state) has a filled level configuration h_u^{10} that is associated with the fully symmetric ^1A_g state. The lowest excited electronic state has the electronic configuration $h_u^9 t_{1u}^1$, i.e., one electron (t_{1u}^1) and one hole (h_u^9). In the group theory, the symmetries for the various states in the $h_u^9 t_{1u}^1$ excitonic state configuration are T_{1g}, T_{2g}, G_g, and H_g obtained by taking the direct product of $T_{1u} \times H_u$. Thus, the optical transitions between the HOMO and the LUMO electronic states are parity-forbidden, since both the electronic states have g-parity. However, with sufficient coupling of the electronic states to intramolecular vibration modes of appropriate symmetry, the oscillator strength can become large enough to activate these parity-forbidden transitions. In other words, each of these four excited state configurations would be combined with appropriate u-parity vibrations to satisfy the electric dipole selection rule and allow coupling to the ^1A_g ground state. This mechanism is called the Hertzberg–Teller mechanism.

Table 11.1 Symmetries of intramolecular vibration modes of C_{60} that induce optical transitions between the ^1A_g electronic ground state and the T_{1g}, T_{2g}, or G_g electronic excited state

State symmetry	Hertzberg–Teller active modes	Jahn–Teller active modes
T_{1g}	a_u, t_{1u}, h_u	h_g
T_{2g}	g_u, h_u	h_g
G_g	t_{2u}, g_u, h_u	g_g, h_g

In Fig. 11.2, the vertical solid lines with arrows up and down represent optical absorption and PL near the absorption edge, respectively. The fine structures observed near the absorption edge in both the optical absorption and the PL spectra of the C_{60} in toluene can be explained with this Hertzberg–Teller vibronic coupling. The symmetries of the intramolecular vibration modes that induce optical transitions between the ground state of ^1A_g symmetry and one of the excited electronic states of T_{1g}, T_{2g}, and G_g symmetries are listed in Table 11.1. The active vibration modes are "u-parity" for the Hertzberg–Teller mechanism and "g-parity" for the Jahn–Teller mechanism. Likewise, similar fine structures are observed near the absorption edge in the optical absorption spectra of solid C_{60} at 80 K. The structures, however, are redshifted by ~40 meV from those of the C_{60} in toluene. This energy difference can be explained by the charge transfer between C_{60} molecules and solvent.

The PL spectra of solid C_{60} in different morphologies such as single crystals, polycrystalline powders, and films have been reported to be different even for nominally similar samples and have been widely known to depend strongly on the crystallinity of the samples. Guss *et al.* interpreted the PL spectra of solid C_{60} as Hertzberg–Teller-induced vibronic transitions originating from C_{60} molecules in a perfect region of the crystal and five different so-called X-traps, which correspond to regions distorted by chemical or physical crystal defects.[24] The chemical impurities can be recognized as residual solvent molecules, residual C_{70}, $C_{60}O_2$, polymerized C_{60}, and so forth. Likewise, the crystal defects are recognized as dislocations, vacancies, and so on. Furthermore, the vibronic structure in the PL spectrum becomes obscured due to strong inhomogeneous broadening in films, compared to single crystals (see Fig. 11.3a, b). Heuvel *et al.* pointed out that the PL spectra of C_{60} single crystals can be explained by two kinds of spectra.[25]

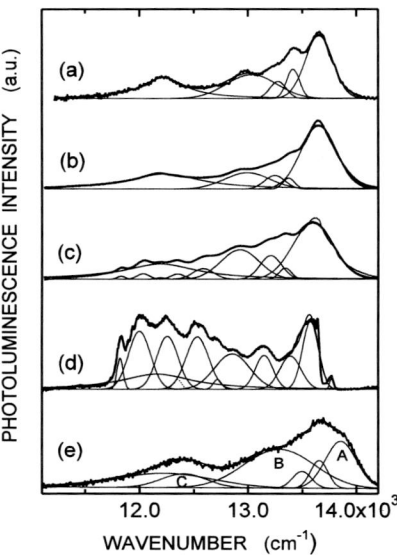

Figure 11.3 PL spectra at 80 K. (a) $C_{60}SC$, (b) $C_{60}TF$ ($t \sim 500$ nm), (c) $C_{60}NC$ ($d \sim 20$ nm), (d) $C_{60}CL$, and (e) $C_{60}NW$ ($d \sim 500$ nm). Each spectrum is decomposed into several appreciable Gaussian and Lorentzian components for guide to eye.

Origins of the two kinds have been discussed in terms of a shallow monomolecular C_{60} trap and a deep trap consisting of a pair of C_{60} molecules in the crystal. The former induces vibronic transitions by the Hertzberg–Teller mechanism and the latter by the Jahn–Teller mechanism. The two PL processes compete as temperature rises. Then, the question is how the vibronic interaction of nanometer-sized C_{60} crystals changes by size effects.

11.3 OPTICAL PROPERTIES OF NANOSCALED PHASES OF SOLID C_{60}

The PL spectra of C_{60} in the form of single crystal (C_{60}SC; 1.5 × 1.5 × 1.0 mm³ in size), thin film (C_{60}TF; about 500 nm in thickness), C_{60}NC (about 20 nm in diameter), cluster (C_{60}CL; composed of less than 100 C_{60} molecules), and C_{60}NW (about 500 nm in diameter) are shown in Fig. 11.3a–e, respectively. C_{60}SC and C_{60}TF were grown in vacuum by using a sublimation method, and C_{60}NC and C_{60}CL were embedded in a SiO matrix by using an inert gas evaporation method.[8] Both vibronic bands derived from the gerade-parity (Jahn–Teller) and ungerade-parity (Hertzberg–Teller) vibration modes appear obviously in nanoscaled phases of C_{60}.

11.3.1 C_{60}NCs

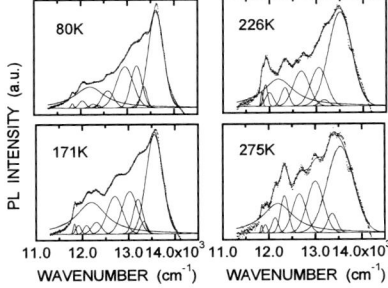

Figure 11.4 Representative PL spectra of C_{60}NC ($d \sim 20$ nm) at 80–275 K. The spectra are decomposed into nine Gaussian components and one Lorentzian component.

Figure 11.4 shows the representative PL spectra of $C_{60}NC$ ($d \sim 20$ nm) at 80, 171, 226, and 275 K. In contrast to the spectra of $C_{60}SC$ and $C_{60}TF$, new vibronic bands appear remarkably in the lower energy region of the PL spectra of $C_{60}NC$. The vibronic bands of $C_{60}NC$ can be assigned on the basis of quantum chemical calculations.[21] Figure 11.5a–c shows the PL spectrum of $C_{60}NC$ at 80 K, the stick diagram of the integrated intensity of the resolved bands, and the oscillator strengths calculated by Negri et al.,[21] respectively.[9] All newly appearing bands can be attributed to the u-parity vibration modes (g_u and t_{2u}). Thus, both vibronic bands derived from the gerade- and ungerade-parity vibration modes obviously appear in the spectrum of $C_{60}NC$. It is well known that above 80 K, the vibronic bands of $C_{60}SC$ and $C_{60}TF$ are associated only with the Jahn–Teller vibronic interaction, and the active modes induced by the interaction are of g-parity. In contrast, those of isolated C_{60}

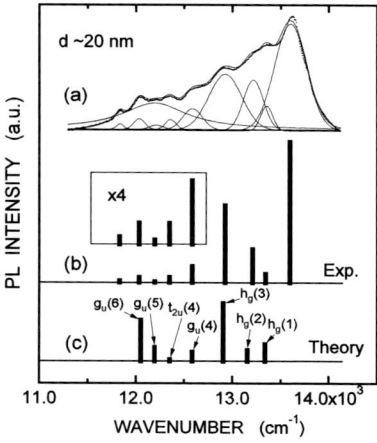

Figure 11.5 (a) PL spectrum of $C_{60}NC$ ($d \sim 20$ nm) at 80 K, (b) stick diagram of the integrated intensity of the resolved bands, and (c) the oscillator strengths calculated by Negri et al.[21] (Reprinted with permission from Ohno et al.,[9] © 2001, American Institute of Physics.)

molecules such as C_{60} in solvents[12] and in rare gas matrices[18] are associated only with the Hertzberg–Teller vibronic interaction, and the active vibration modes are of u-parity. The results on $C_{60}NC$, therefore, indicate that the Hertzberg–Teller-type transitions become strong with a decrease in the size of C_{60} crystal.[9]

By plotting the integrated intensity of each vibronic band, the Hertzberg–Teller-type vibronic bands are found to become relatively dominant with an increase in temperature, compared to the Jahn–Teller-type bands. Analyses on the temperature dependence of the quantum efficiency indicate that the activation energy for the nonradiative relaxation can be classified into two groups: 190 and 90 meV for the Hertzberg–Teller- and Jahn–Teller-type vibronic bands, respectively.[9] The results indicate that two kinds of vibronic origins coexist in C_{60}NC.

11.3.2 C_{60}NWs and C_{60}NTs

The PL spectrum of C_{60}NW in Fig. 11.3e exhibits a spectral feature broader than that of C_{60}SC and looks similar to that of C_{60}TF at a glance. However, appreciable differences between C_{60}NW and C_{60}TF are seen. As shown in Fig. 11.3e, three remarkable PL components denoted as A, B, and C appear in C_{60}NW. Consequently, the PL spectrum of C_{60}NW looks shifted overall to a higher energy as compared with that of C_{60}TF. Even in C_{60}SC, an apparent diversity of PL spectra has been observed in the low temperature phase below 90 K. Two types of PL spectra of pristine C_{60} crystals were reported at low temperatures around 5 K.[25,26] The superposition of the two contributions with the relative intensity varied from sample to sample should be responsible for the diversity in PL spectra. The coexistence of the two types of PL spectra originates from inhomogeneously broadened localized states caused by intrinsic defect or lattice disorder in the low temperature phase.[26] Thus, the inhomogeneity in excited states due to defects or surface may be responsible for the broadening and the diversity in PL spectra observed in C_{60}NW and C_{60}NT. It should be noted that the three PL components denoted by A, B, and C in Fig. 11.3e correspond roughly to the broadened spectrum of type B in Ref. 26, which disappears rapidly above 25 K for pristine C_{60} crystals.

To confirm the inhomogeneity as an intrinsic characteristic for C_{60}NW and C_{60}NT, micro-PL spectra were measured.[27] Figure 11.6

shows a scanning electron microscopy (SEM) image of the structurally improved C_{60}NT sample by methanol soaking followed

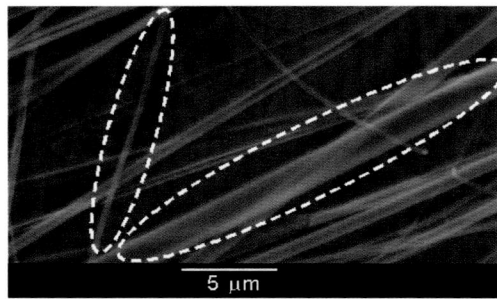

Figure 11.6 SEM image of the structurally improved C_{60}NT sample by methanol soaking followed by annealing at 260°C for 5 h. A nanotube and a microtube with the outer diameters of about 500 nm and 2.5 μm, respectively, are seen as circled by dashed lines.

by annealing at 260°C for 5 h.[28] Besides C_{60}NT as small as 500 nm in outer diameter, microtubes with a couple of microns in outer diameter are seen in the SEM image. Figure 11.7 shows the micro-PL spectra of C_{60}NT (about 500 nm in outer diameter) and C_{60} microtube (2.5 μm in outer diameter) at two different measuring spots each at 4 K. The micro-PL spectra of the C_{60}NT exhibit broader features than those of the microtube and are varied by changing the measuring spot in the nanotube. This suggests that the broad and spot-sensitive PL spectra observed for the C_{60}NT is characteristic of the one-dimensional submicron-scaled C_{60} crystals. Figure 11.8 shows that the presence of residual solvent molecules in C_{60}NW broadens the PL spectrum and induces a higher energy component. The as-grown sample before annealing exhibits a broad X-ray diffraction profile due to poor crystallinity and the presence of residual solvents of both *m*-xylene and isopropyl alcohol molecules by Fourier transform infrared spectroscopy measurements. After annealing at 250°C for 5 h, *m*-xylene molecules are removed from the sample and thermal structural relaxation occurs, resulting in the sharper PL spectrum. Thus, PL spectra of C_{60}NW and C_{60}NT exhibit the size dependence related sensitively to the crystallinity associated with intrinsic defects due to the large surface-to-volume ratio or the presence of residual solvents.

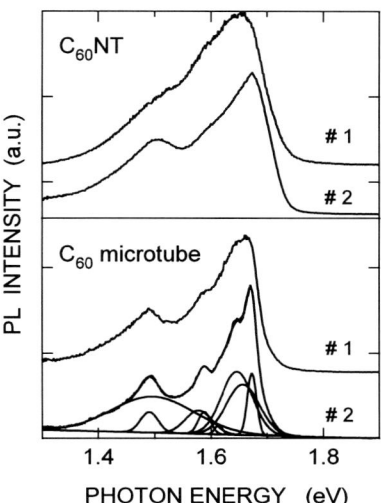

Figure 11.7 Micro-PL spectra of C_{60}NT and C_{60} microtube at two different measuring spots #1 and #2 each at 4 K. The outer diameters of C_{60}NT and C_{60} microtube are about 500 nm and 2.5 μm, as shown by the dashed circles in Fig. 11.6. The lowermost spectrum is decomposed into several appreciable Gaussian components for guide to eye.

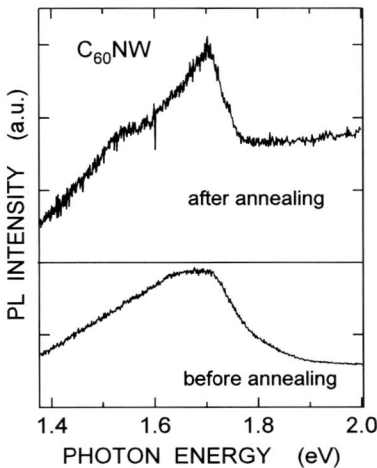

Figure 11.8 PL spectra of C_{60}NW at room temperature before and after annealing at 250°C for 5 h.

11.4 PHOTO-INDUCED STRUCTURAL TRANSFORMATIONS OF C_{60}NWs AND C_{60}NTs

Photo-induced structural transformations of C_{60} crystals to a polymerized phase and to a glassy graphitic phase also depend strongly on its morphology, crystallinity, and environment (temperature, pressure, atmosphere, etc.).[29–32] The Raman scattering spectra of C_{60}SC, C_{60}NW, and C_{60}NT under laser irradiation by the 514.5 nm line from an Ar⁺ laser were measured to investigate the effects of size and shape on the photo-induced polymerization and graphitization kinetics.[27] Figure 11.9a shows the time evolution of the Raman spectrum of C_{60}NW (annealed at 220°C for 5 h) under the laser irradiation of 75 W/cm² at room temperature in vacuum.

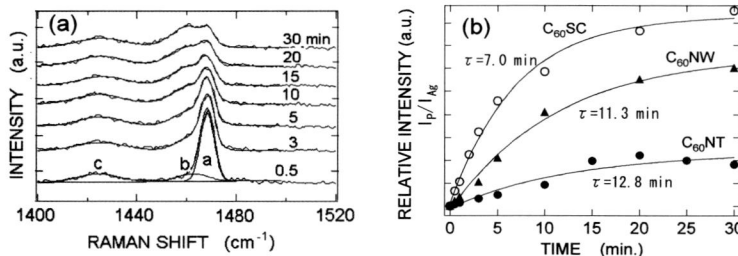

Figure 11.9 (a) Change in the Raman spectrum of C_{60}NW with laser irradiation time (514.5 nm, 75 W/cm²). Experiment was carried out at room temperature in vacuum. (b) Relative Raman intensity of polymer mode to $a_g(2)$ pentagonal pinch mode as a function of laser irradiation time for C_{60}SC (open circles), C_{60}NW (solid triangles), and C_{60}NT (solid circles). Both C_{60}NW and C_{60}NT were annealed at 220°C for 5 h before the laser irradiation.

The changes in the Raman spectra are irreversible and ascribed to the photopolymerization occurring in C_{60}NW. The Raman spectra in this frequency range can be decomposed into three Gaussian components, as indicated by a, b, and c. Modes a and c are assigned to the $a_g(2)$ pentagonal pinch mode and the h_g mode of C_{60} molecular vibrations, respectively.

The appearance of mode b in the vicinity of the $a_g(2)$ mode is attributed to the formation of interconnected C_{60} molecules as dimers or polymeric chains. Hereafter, mode b is referred to as the polymer mode. The polymer mode increases in intensity and decreases in frequency from 1465 to 1458 cm^{-1} as the laser irradiation proceeds. The softening of the polymer mode can be attributed to the decrease of the number of double bonds in a C_{60} molecule by the [2+2] cycloaddition of adjacent molecules with polymerization. Figure 11.9b shows the relative intensity of the polymer mode to the $a_g(2)$ pentagonal pinch mode as a function of laser irradiation time for C_{60}NW, C_{60}NT, and C_{60}SC. The photopolymerization proceeds with the time constants τ = 7.0, 11.3, and 12.8 min for C_{60}SC, C_{60}NW, and C_{60}NT, respectively. The observation indicates that the photopolymerization is suppressed in C_{60}NW and C_{60}NT compared with C_{60}SC. In addition, the photopolymerization for C_{60}NW proceeds faster after annealing than before.[27] The improvement of crystallinity in C_{60}NT promotes photopolymerization.[28] Thus, the poorer crystallinity as well as the presence of solvent molecules in C_{60}NW and C_{60}NT would suppress the polymerization kinetically and structurally compared with C_{60}SC.

When the samples are irradiated in air with the same laser power, graphitization is observed instead of polymerization. The presence of oxygen molecules in atmosphere induces the destruction of C_{60} molecules by the photoexcitation of C_{60}-O_2, following the production of a carbonyl-like bonding configuration. This process results in the graphitization of solid C_{60}. Figure 11.10a shows the time evolution of the Raman spectrum (in 1300–1700 cm^{-1}) of C_{60}NW under the laser irradiation of 75 W/cm^2 at room temperature in air. As the laser irradiation time increases,

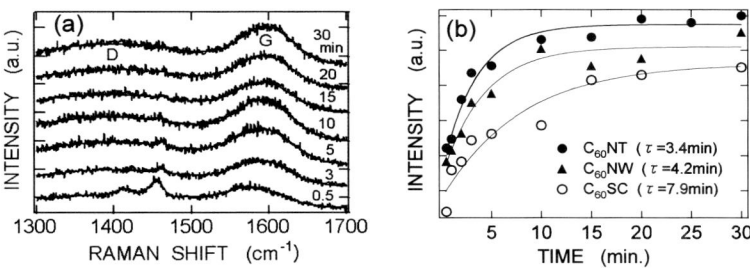

Figure 11.10 (a) Change in the Raman spectrum of C_{60}NW with laser irradiation time (514.5 nm, 75 W/cm^2). Experiment was carried out at room temperature in air. (b) Intensity of G band as a function of laser irradiation time for C_{60}SC (open circles), C_{60}NW (solid triangles), and C_{60}NT (solid circles). Both C_{60}NW and C_{60}NT were annealed at 220°C for 5 h.

the $a_g(2)$ pentagonal pinch mode at 1467 cm^{-1} decreases in intensity and two broad bands appear at about 1400 and 1590 cm^{-1}. Those broad bands grow and shift slightly to higher frequencies with an increase in the laser irradiation time. The two broad bands, which are sometimes referred to as D and G bands, respectively, are characteristic of disordered sp^2 carbon (i.e., glassy graphitic carbon). Figure 11.10b shows an increase in the intensity of the G band with laser irradiation time for C$_{60}$NW, C$_{60}$NT, and C$_{60}$SC. The photo-induced graphitization proceeds with the time constants τ = 7.9, 4.2, and 3.4 min for C$_{60}$SC, C$_{60}$NW, and C$_{60}$NT, respectively. Thus, the graphitization is promoted in C$_{60}$NW and C$_{60}$NT, compared with C$_{60}$SC. Since the surface area is significantly larger for C$_{60}$NT and C$_{60}$NW than C$_{60}$SC, the graphitization can be promoted in the aggregated nanoscaled C$_{60}$. By comparing the time constant of the kinetics before and after annealing for C$_{60}$NW, it is found that the presence of residual solvent molecules in samples can also promote the graphitization.[27] Thus, photo-induced polymerization and graphitization in C$_{60}$NW and C$_{60}$NT are affected by their crystallinity, sample environment, and the presence of residual solvent molecules.

11.5 CONCLUDING REMARKS

The PL spectra of C$_{60}$NW and C$_{60}$NT were compared with those of C$_{60}$SC, C$_{60}$TF, C$_{60}$NC, and C$_{60}$CL. The micro-PL spectra of C$_{60}$NW and C$_{60}$NT exhibit broad PL components and depend on the local positions of samples. The observations suggest that the inhomogeneity in excited states exists due to defects or surface states for C$_{60}$NW and C$_{60}$NT. The presence of residual solvent molecules also influences PL spectra. The photo-induced polymerization and graphitization kinetics were also discussed for C$_{60}$NW and C$_{60}$NT. The photopolymerization is suppressed, while the photo-induced graphitization is promoted in C$_{60}$NW and C$_{60}$NT compared with C$_{60}$SC. Those photo-induced structural transformations are affected not only by the sample size but also by their crystallinity, sample environment, and the presence of residual solvent molecules.

References

1. T. P. Martin, U. Naher, H. Schaber, and U. Zimmermann, *Phys. Rev. Lett.*, **70**, 3079 (1993).
2. J. S. Ahn, K. Suzuki, Y. Iwasa, and T. Mitani, *J. Lumin.*, **72–74**, 464 (1997).
3. J. S. Ahn, K. Suzuki, Y. Iwasa, N. Otsuka, and T. Mitani, *J. Lumin.*, **76–77**, 201 (1998).
4. G. Gu, W. Ding, G. Cheng, W. Zang, H. Zen, and Y. Du, *Appl. Phys. Lett.*, **67**, 326 (1995).
5. G. Gu, W. Ding, Y. Du, H. Huang, and S. Yang, *Appl. Phys. Lett.*, **70**, 2619 (1997).
6. G. Gu, W. Ding, G. Cheng, S. Zhang, Y. Du, and S. Yang, *Chem. Phys. Lett.*, **270**, 135 (1997).
7. S. Y. Wang, W. Z. Shen, X. C. Shen, L. Zhu, Z. M. Ren, Y. F. Li, and K. F. Liu, *Appl. Phys. Lett.*, **67**, 783 (1995).
8. T. Ohno, K. Matsuishi, and S. Onari, *J. Appl. Phys.*, **83**, 4939 (1998).
9. T. Ohno, K. Matsuishi, and S. Onari, *J. Chem. Phys.*, **114**, 9633 (2001).
10. K. Miyazawa, Y. Wuwasaki, A. Obayashi, and M. Kuwabara, *J. Mater. Res.*, **17**, 83 (2002).
11. J. Minato, K. Miyazawa, and T. Suga, *Sci. Technol. Adv. Mater.*, **6**, 272 (2005).
12. Y. Wang, *J. Phys. Chem.*, **96**, 764 (1992).
13. Y. Zeng, L. Biczok, and H. Linschitz, *J. Phys. Chem.*, **96**, 5237 (1992).
14. K. Palewska, J. Sworakowski, H. Chojnacki, E. C. Meister, and U. P. Wild, *J. Phys. Chem.*, **97**, 12167 (1993).
15. D. J. van den Heuvel, I. Y. Chen, E. J. J. Groenen, J. Schmidt, and G. Meijer, *Chem. Phys. Lett.*, **231**, 111 (1994).
16. D. J. van den Heuvel, G. J. B. van den Berg, E. J. J. Groenen, J. Schmidt, I. Holleman, and G. Meijer, *J. Phys. Chem.*, **99**, 11644 (1995).
17. A. Sassara, G. Zerza, and M. Chergui, *Chem. Phys. Lett.*, **261**, 213 (1996).
18. A. Sassara, G. Zerza, M. Chergui, F. Negri, and G. Orlandi, *J. Chem. Phys.*, **107**, 8731 (1997).

19. A. Sassara, G. Zerza, and M. Chergui, *J. Phys. B: At. Mol. Opt. Phys.*, **29**, 4997 (1996).
20. W. C. Hung, C. D. Ho, C. P. Liu, and Y. P. Lee, *J. Phys. Chem.*, **100**, 3927 (1996).
21. F. Negri, G. Orlandi, and F. Zerbetto, *J. Chem. Phys.*, **97**, 6496 (1992).
22. F. Negri, G. Orlandi, and F. Zerbetto, *J. Phys. Chem.*, **100**, 10849 (1996).
23. M. S. Dresselhaus, G. Dresselhaus, and P. C. Eklund, *J. Mater. Res.*, **8**, 2059 (1993).
24. W. Guss, J. Feldmann, E. O. Göbel, C. Taliani, H. Mohn, W. Müller, P. Häussler, and H. U. ter Meer, *Phys. Rev. Lett.*, **72**, 2644 (1994).
25. D. J. van den Heuvel, I. Y. Chen, E. J. J. Groenen, M. Matsushita, J. Schmidt, and G. Meijer, *Chem. Phys. Lett.*, **233**, 284 (1995).
26. I. Akimono and K. Kan'no, *J. Phys. Soc. Jpn.*, **71**, 630 (2002).
27. K. Matsuishi, K. Minamiru, and K. Naito, unpublished manuscript.
28. K. Naito and K. Matsuishi, *J. Phys.: Conf. Ser.*, **159**, 012020 (2009).
29. K. Matsuishi, K. Tada, S. Onari, T. Arai, R. L. Meng, and C. W. Chu, *Phil. Mag. B*, **70**, 795 (1994).
30. K. Matsuishi, T. Ohno, N. Yasuda, T. Nakanishi, S. Onari, and T. Arai, *J. Phys. Chem. Solids*, **58**, 1747 (1997).
31. T. Ohno, K. Matsuishi, and S. Onari, *Solid State Commun.*, **101**, 785 (1997).
32. T. Mitani, K. Matsuishi, and S. Onari, *Chem. Phys. Lett.*, **383**, 486 (2004).

Chapter 12

SURFACE NANOCHARACTERIZATION OF FULLERENE NANOWHISKERS

Daisuke Fujita and Mingsheng Xu

National Institute for Materials Science, 1-2-2 Sengen, Tsukuba, Ibaraki 305-0047, Japan
fujita.daisuke@nims.go.jp

The surface characterization methodology for the fullerene nanomaterials (FNMs) is overviewed with a strong emphasis on the uses of a variety of scanning probe microscopy (SPM). Among numerous SPM techniques, atomic force microscopy (AFM) and its major variants are introduced and the related applications to FNM studies are discussed. Starting from the introduction of scanning force microscopy, several modes of the AFM operation, such as contact mode, intermittent contact mode, noncontact mode, frictional force imaging, phase imaging, conductive measurement mode, and so on, are shown. The most common and important use of AFM is the three-dimensional topography imaging of the nano-objects at the nanoscale. However, the raw AFM topography images always involve various artifacts. The most common artifact is caused by the finite dimensions of probe tip. Precise and quantitative morphology analysis of the FNMs can be realized by the use of proper

Fullerene Nanowhiskers
Edited by Kun'ichi Miyazawa
Copyright © 2012 Pan Stanford Publishing Pte. Ltd.
www.panstanford.com

image reconstruction with actual probe shape functions. Finally, the thermal stability of FNMs investigated by a variable temperature SPM at elevated temperatures is discussed.

12.1 INTRODUCTION

This chapter deals with surface characterization techniques of fullerene nanowhiskers (FNWs), fullerene nanotubes (FNTs), and other related fullerene nanomaterials (FNMs) at the nanometer scale. In the last decades, both scanning electron microscopy (SEM) and transmission electron microscopy (TEM) have been generally used for the nanoscale visualization of shapes and sizes of FNMs.[1,2] However, these electron microscopy techniques using relatively high-energy electron beams are not considered as surface-sensitive analytical methods.

Here we discuss the surface-sensitive analytical techniques with nanometer-scale resolution. By the way, as for the mesoscopic surface analytical tools, scanning Auger microscopy (SAM) with a field-emission (FE) electron gun is a typical workhorse. While a finely focused beam is scanned over a sample surface, the emitted Auger electrons are measured for the chemical mapping of sample surfaces at the special resolution of around 10–100 nm. Since the inelastic mean free paths of Auger electrons are generally less than a few nanometer, FE-SAM is very sensitive to the surface especially in the case of low-energy Auger electrons.[3,4]

However, analytical microscopy with the highest surface sensitivity and spatial resolution is a family of scanning probe microscopy (SPM). More than two decades have passed since the invention of scanning tunneling microscopy (STM) and atomic force microscopy (AFM).[5,6] STM was invented by Binnig and Rohrer in 1982, which was based on a quantum tunneling effect and thus it is applicable only to conductive surfaces.[5] When a sharp conductive tip or a probe is brought into a close proximity (<1 nm) with a sample surface, the application of a DC bias voltage between the tip and the sample may allow electrons to tunnel through the potential barrier. Quantum tunneling is a short-range interaction and the tunneling current decays exponentially with the distance. Thus, the tunneling current is very suitable for the feedback signal source of the gap-distance control at the picometer resolution. Therefore, STM is suitable to probe the electronic structure of conductive surfaces at the subnanometer scale. The application of STM to fullerene researches has been limited to molecular adsorption or initial stage of ultrathin film growth.[7–9] There has been few research on bulky FNMs up to now due to their low conductivity after exposed to the air.

After the invention of STM, numerous types of scanned probe microscopies have been invented, where various quantities and properties can be measured through a variety of tip–surface interactions. Among the SPM family, the most widely used microscopy is scanning force microscopy (SFM) or AFM, which was invented in 1986 by Binnig *et al.*.[6] Feedback signal sources of AFM are various types of short- and long-range forces interacting between a probe tip and a sample surface. Since the AFM and its derivatives do not require the electric conductivity for a probe tip and a sample surface, AFM-based SPMs can be applied not only to conductive materials but also to insulating or poorly conductive materials such as organic materials including fullerene-based nanomaterials. Therefore, in this chapter, recent development of surface nanocharacterization on the FNMs is reviewed mainly from the viewpoints of AFM-based SPMs.

12.2 CONCEPTS OF AFM

Nowadays, dominant majority of SPMs are derivatives of AFM, where the quantitative measurement of forces between a probe tip and a sample surface is the key technology. Although various types of force sensors have been proposed so far, the most common type of force sensors is based on the quantitative detection of the deflection of a cantilever.

Normally, a probe tip with a very sharp protrusion is mounted at the end of a cantilever. Different measurement techniques have been developed for the detection of a minute bending of a cantilever, in which electron tunneling,[6] optical reflection,[10] interferometry,[11] capacitance,[12] piezoresistance,[13] and so on, are involved as the signals. Among them, the most popular detection method is the so-called beam-deflection method, in which a laser-light beam is reflected at the backside of a cantilever and impinges on a position-sensitive photon detector.

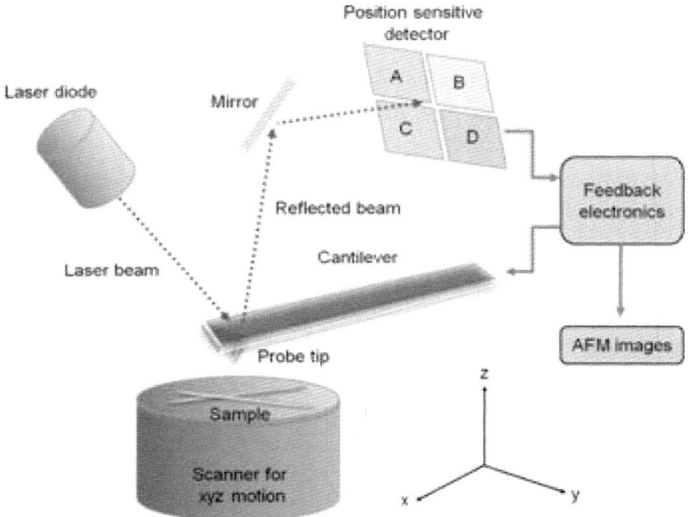

Figure 12.1 A schematic illustration of the operating principle of conventional AFM of beam-deflection type. The sample is mounted on an *xyz*-motion scanner. A sharp probe tip at the end of a cantilever is positioned near the surface of a sample with a coarse positioning device. Normal or lateral force acting on the tip apex can be measured by normal or torsional deflection of the cantilever. A laser beam is reflected off the rear side of the cantilever. The deflections of the cantilever are detected with a position-sensitive detector made of a four-segment photodiode detector.

A schematic illustration of a conventional beam-deflection AFM is shown in Fig. 12.1. Normal or lateral force acting on the tip apex can be measured by normal and/or torsional deflection of a cantilever. Either cantilever deflection may cause a change in the optical path of a reflected beam, which can be quantitatively detectable by a position-sensitive four-segment photodiode detector. For example, the signal of $Z = (A + C) - (B + D)$ is proportional to the normal force, whereas that of $X = (A + B) - (C + D)$ is proportional to the lateral force. A sharp probe tip at the end of a cantilever is positioned near the surface of a sample with a coarse positioning device. By using a feedback signal such as Z, proportional to forces acting between a tip and a surface, the gap distance is maintained constant. By scanning a sample or a tip, a variety of tip–surface interactions can be imaged by AFM, depending on the gap distance. There are several operation modes of AFM measurements, which can be categorized in static and dynamic modes.

12.3 STATIC MODE AFM IMAGING

In the static mode AFM imaging, the static deflection of a cantilever tip is used as a feedback signal. The static mode AFM can produce an image of a topographic image of a sample surface measured at a constant normal force. When a probe tip is approached very close to a surface, attractive forces may become strong enough to cause the tip to "jump-to-contact" with the surface. Thus the static AFM is almost always operated in contact with the surface where the tip experiences a repulsive force with a typical value of 10^{-9} N. Consequently, the static AFM is generally called "contact mode AFM."

In the contact mode, a force between a tip and a sample causes the cantilever to deflect in accordance with Hooke's law. Thus the normal force applied by an external cantilever can be calculated by using the normal spring constant of the cantilever. Although the contact AFM is operated in the repulsive force regime, it should be noted that the total force is composed of the short-range repulsive and long-range attractive forces. In general, the contact AFM is operated in ambient atmosphere. Although atomic features can be imaged on the surfaces of highly oriented pyrolytic graphite or hexagonal boron nitride,[14,15] those are limited to lattice imaging. True atomic resolution in the contact mode has not been reported so far. In principle, the spatial resolution of a contact AFM is limited by the contact area.

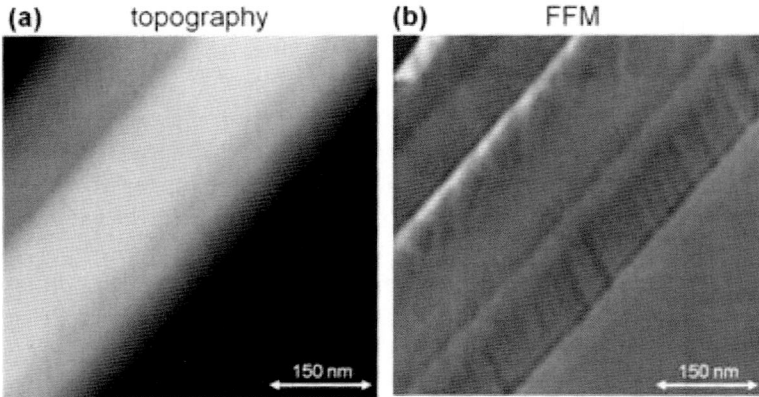

Figure 12.2 (a) A topography image of an FNW measured simultaneously in contact mode AFM in air. (b) An FFM image of the FNW simultaneously measured with (a).

When the cantilever tip is scanned perpendicular to its length in the contact mode, a nanoscopic friction in the contact area produces a lateral force on the tip apex. This lateral force or friction force causes a torsional deflection of the cantilever. The measurement technique of lateral force imaging is called lateral force microscopy (LFM) or friction force microscopy (FFM).[16] Using a four-segment photodiode detector as shown in Fig. 12.1, topography and FFM images can be simultaneously measured in the contact mode. Figure 12.2 shows topography and friction force images measured on FNWs in contact mode AFM in ambient atmosphere. Contrast of lateral friction force is observed on the top and side walls, suggesting the presence of geometrical or chemical inhomogeneity on the surfaces of FNWs.

12.4 DYNAMIC MODE AFM IMAGING

In dynamic mode AFM, the vibrational properties of the cantilever related to the tip–sample interaction forces are measured while the cantilever tip is externally oscillated at or close to its resonance or harmonic frequency. There are several techniques in the category of dynamic AFM.[17] The oscillation amplitude, resonance frequency, and phase between excitation and oscillation of the cantilever are modified by tip–sample interaction forces. The dynamic mode operation can be categorized into frequency modulation (FM) and amplitude modulation (AM). In the FM mode of operation, the shift in resonance frequency caused by the tip–sample interaction force is used as the feedback signal for AFM imaging. This mode is also called "noncontact AFM" or NCAFM. The NCAFM with FM scheme was the first reliable technique to demonstrate true atomic resolution on a Si(111) reconstructed surface in ultrahigh vacuum.[18]

In the AM mode of operation, the change in the amplitude or phase of an oscillation excited at a fixed frequency close to the resonance frequency is used as the feedback signal for AFM imaging. Amplitude modulation can be operated either in the noncontact regime with an attractive force or in the intermittent contact regime with a repulsive force. In the case of intermittent contact or the so-called tapping mode, a sinusoidal oscillation of the tip with the oscillation amplitude of 20–100 nm is excited at a little below the resonance frequency. The intensity of the oscillation amplitude is used for the feedback source signal.[19] The decrease in the oscillation amplitude may be due either to the intermittent contact or tapping the surface or to the shift in resonance frequency. AM can also be used in the noncontact regime where the cantilever is oscillated at a frequency slightly above its resonance frequency with the oscillation amplitude of a few nanometers. Attractive van der Waals forces or the so-called dispersion forces act to decrease the resonance frequency, which also decreases the oscillation amplitude.

In the dynamic mode using a lock-in amplifier, the phase shift between the driving oscillator and the oscillation cantilever can be obtained and utilized for simultaneous imaging with topography mapping. This type of imaging called "phase imaging" can be related to the stiffness of the surface under certain conditions.[20] Generally, the phase imaging is used for the visualization of chemical inhomogeneity on the surface.

Typical height and phase images of an FNW observed by intermittent contact mode AFM are shown in Fig. 12.3. The height image of FNW clearly indicates that the surface is relatively rough and composed of nanoscale grains. The phase imaging confirms that those nanoscale grains have different contrast, suggesting inhomogeneity in stiffness.

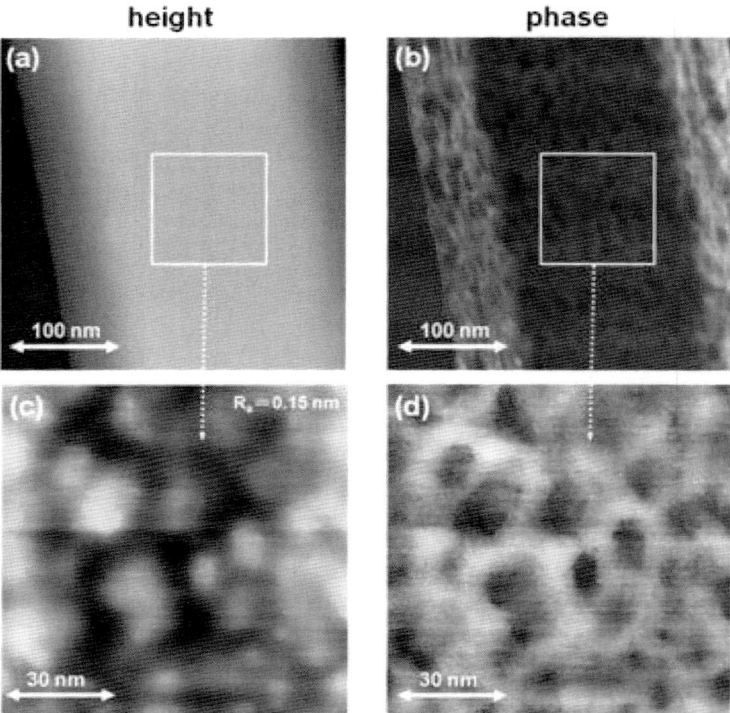

Figure 12.3 An FNW imaged by intermittent contact mode AFM. (a) AFM height image. (b) The simultaneously obtained phase image. (c) The enlarged height image on the top surface of FNW. (d) The simultaneously obtained phase image. Average roughness R_a = 0.15 nm. Peak-to-valley roughness = 2.5 nm.

Figure 12.4 shows the typical height, amplitude, and phase images of an FNW dispersed on a Si(001) substrate observed by NCAFM with AM detection. Not only an FNW but also fullerene nanoparticles are found on the substrate surface. Both the amplitude and the phase images clearly show the contrast distribution on the surface of the FNW, which indicates the presence of chemical inhomogeneity.

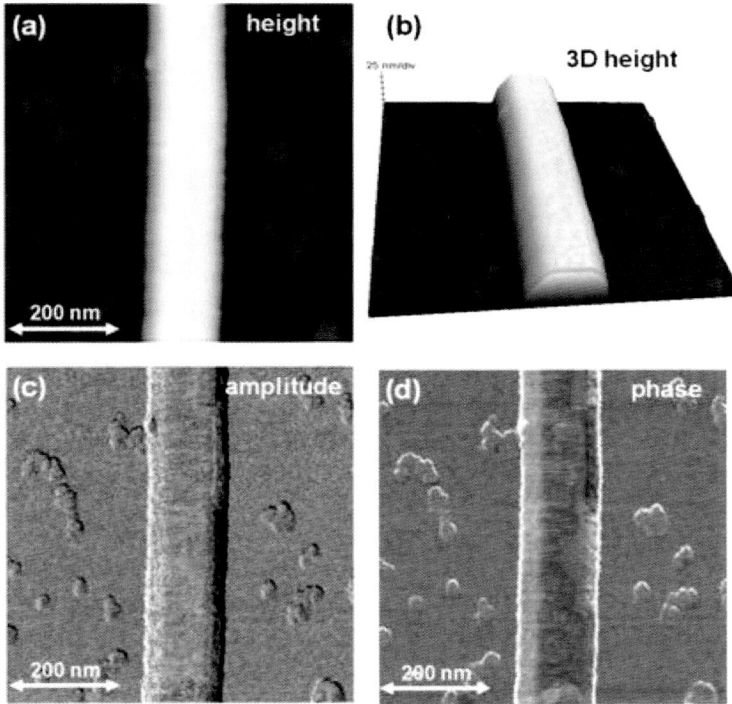

Figure 12.4 A single FNW and fullerene nanoparticles dispersed on a Si(001) substrate imaged by NCAFM. (a) NCAFM height image. (b) Three-dimensional height image of (a). (c) NCAFM amplitude image. (d) NCAFM phase image.

12.5 RESTORATION OF IMAGE ARTIFACTS

In order to promote the societal acceptance of FNMs, it is required to specify the properties using unified terminology and methodology. Especially, quantitative characterization methodology at the nanometer-scale shall be established. Thus, it is important to standardize nanocharacterization methodology of AFM based on the authorized schemes for international standardization.[21,22]

Obviously, the most significant role of AFM is the precise measurement of dimensions and morphology. However, it should be noted that AFM imaging involves intrinsic or extrinsic artifacts. The extrinsic artifacts are caused by drift, noise, vibration, insufficient optimization of set parameters, calibration errors, and so on. The intrinsic artifact of AFM imaging is caused by the finite size of probing tips. For quantitative reduction of the tip artifact, image restoration methodology for distorted images shall be established.[23,24]

AFM images are created by scanning a sharp tip over a sample surface and recording relevant signals. Because the probing tip has its own shape with a finite size, the acquired image does not reflect the actual specimen shape. The acquired height image is formed by the so-called dilation process between the probing tip shape and the sample topography feature. It should be noted that significant distortion in topography imaging may occur if the sample surface has relatively large corrugation compared to the dimensions of the probing tip.[25]

Based on the mathematical morphology, the real surface topography, $s(x, y)$, is dilated to the measured height image representing the apparent surface topography, $z(x, y)$, by the finite probe-tip shape, $p(x, y)$:

$$z(x, y) = \max\{s(x', y') - p(x'-x, y'-y)\} \quad (12.1)$$

The acquired height image through the dilation process can be partially restored by using the so-called erosion process which can reconstruct the upper bound image, $r(x, y)$, of the actual surface topography:

$$r(x, y) = \min\{z(x', y') + t(x'-x, y'-y)\} \quad (12.2)$$

The probe shape function, $p(x, y)$, can be deduced by using tip characterizers that may be calibrated to ensure a certain dimensional precision.[26] The image restoration process is applied to $z(x, y)$ by using $p(x, y)$ to extract the reconstructed surface, $r(x, y)$.

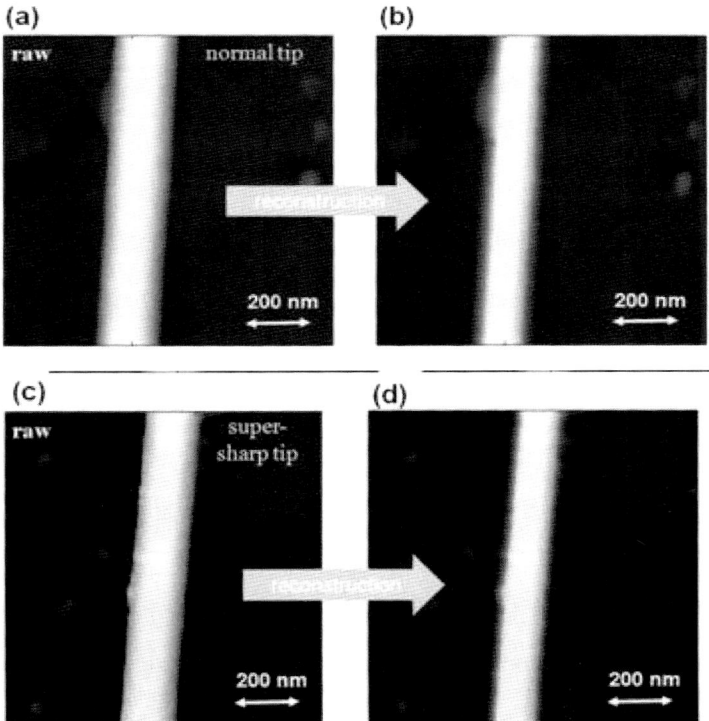

Figure 12.5 (a) Intermittent contact AFM image of an FNW using a normal tip with a curvature radius of 10 nm. (b) The reconstructed image of (a). (c) A height image of an FNW using a supersharp tip with a curvature radius of 2 nm. (d) The reconstructed image of (c).

The image reconstruction procedure using an actual probe shape function is applied to intermittent contact mode images of an FNW, as shown in Fig. 12.5. The AFM images were taken with two different probes with different tip curvatures. Although the supersharp tip gives relatively sharper image in the raw data, both of the reconstructed images show a similar width and height. The deduced ratios between the half width at half maximum and height obtained from the reconstructed images are almost the same, which indicates that the reconstruction is reliable. Furthermore, the ratios agree very much with that estimated from the typical FNWs with hexagonal cross sections.[27]

12.6 CONDUCTIVE AFM WITH FE-SAM

The C_{60} fullerene exhibits semiconducting properties and the electrical properties can be tailored.[28,29] The C_{60} fullerene thin films are used as active materials in n-type organic field-effect transistors (FETs) with a relatively high mobility of ~1.0 cm²/(V s) as well as in solar cells.[30,31] Since the one-dimensional structures of FNWs may be of potential importance for nanoelectronics, the electronic and transport properties of FNWs have been investigated using a variety of measurement devices with C_{60} FNW-FET,[32] contacting two-point probe,[33,34] and contacting four-point probe.[35] Those electric measurements are conventional methodology operated at the micron scale. Recently, the electrical conductivity of individual C_{60} FNWs at the nanoscale has been clarified by exploiting the so-called conductive AFM.[36]

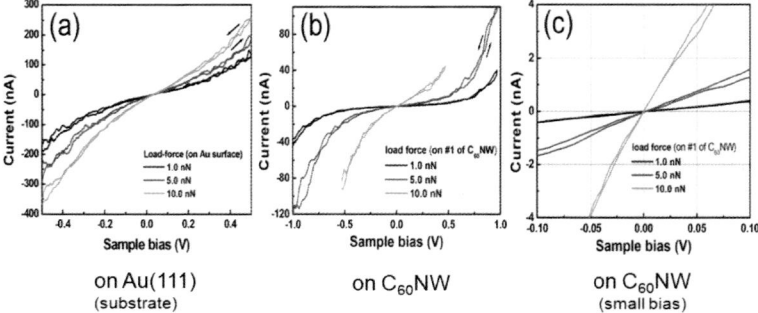

Figure 12.6 (a) *I–V* spectra acquired on (a) Au(111) substrate, (b) an FNW of ~100 nm diameter, and (c) the same FNW with a small bias range. PtIr-coated Si cantilever probe is used for conductive AFM mapping. Point *I–V* spectroscopy was performed at the specified positions with different contact forces ranging from 1.0 to 10.0 nN.

The conductive AFM is a variant of contact mode AFM using a conductive probe tip by which the variations in surface conductivity can be distinguished. While a DC bias is applied to the tip or sample, current passing between the tip and the sample is measured to generate a conductive AFM map as well as a corresponding topography image. It is also possible to measure current–voltage characteristics (*I–V* spectroscopy) at specified positions in an AFM image.

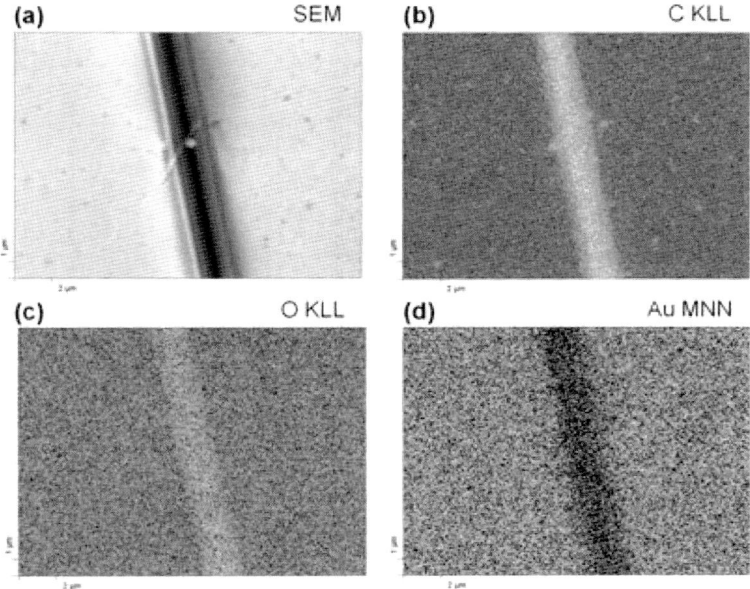

Figure 12.7 FE-SAM analysis of the C_{60} FNWs on the Au(111) substrate. (a) FE-SEM image. (b–d) Auger maps of C KLL, O KLL, and Au MNN.

Typical current–voltage characteristics of a single FNW (~100 nm in diameter) dispersed on an Au(111) substrate are shown in Fig. 12.6. The pristine C_{60} FNW as received is conducting, and its conductivity increases with the increase in the contact force. However, the conductivity of the FNWs exposed to ambient air is found to decrease with the elapsed time and finally becomes insulating.

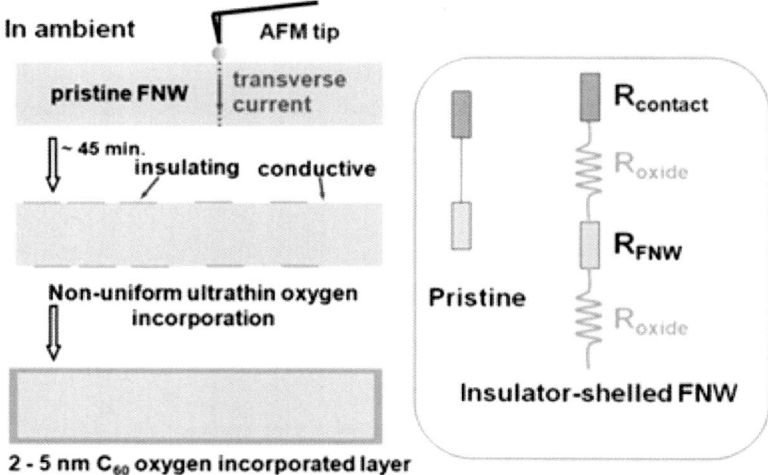

Figure 12.8 A schematic illustration of the formation of oxygen intercalation layer at an FNW exposed to ambient atmosphere. At the beginning, oxygen adsorption occurs randomly and creates less conductive areas at the surface. Finally, the oxygen incorporated layer of several nanometer thicknesses is formed on the entire surface of the FNW, which means the creation of an insulator-shelled FNW.

Auger electron microscopy with an FE electron gun (FE-SAM) enables us to clarify the surface chemical analysis of the FNWs exposed to the air. The state-of-the-art FE-SAM can measure FE-SEM images as well as Auger maps with the spatial resolution of ~10 nm. Figure 12.7 shows typical SEM and SAM images observed on the FNWs on an Au(111) substrate. It should be noted that the spatial distribution of C KLL and O KLL shows a good accordance, which suggests that the surface of the air-exposed FNWs involves a few atomic percent of oxygen. The observed oxygen intercalation layer can be removed by Ar^+ ion sputtering easily. The estimated thickness of the oxygen intercalation layer corresponds to a few monolayers of C_{60} in its pure solid form with a face-centered cubic lattice. It has been reported that the exposure to oxygen induces adsorption and diffusion of oxygen, which results in an abrupt decrease of the conductivity and mobility of C_{60} films by several orders of magnitude.[37,38] The negative effects of the oxygen intercalation on the conductivity can be understood in terms of the creation of deep electronic states that causes the Fermi level to shift down to the middle of the gap. The state acts as an efficient trap for electrons in the conduction band and as nonradiative recombination centers.[39]

The insulating-layer formation process promoted by oxygen incorporation into the FNWs is schematically shown in Fig. 12.8. The outer oxygen-enriched layer may play a role like that of SiO_2 in the Si industry. It may be used as a self-formed insulating shield without an additional dielectric layer deposition, which can be used in FNW-based FETs for future nanoelectronics.

12.7 HIGH-TEMPERATURE AFM ANALYSIS

The thermal stability of FNMs is a matter of interest for the actual device applications. Recently, the structural change of FNTs at 180°C in ambient atmosphere has been reported.[40] The in situ analysis of the morphology of FNMs at the elevated temperature can be realized by using a variable temperature scanning probe microscopy (VT-SPM). Compared with high-temperature STM in ultrahigh vacuum (UHV), the number of high-temperature AFM studies is still limited.[41,42]

Figure 12.9 shows an NCAFM image of fullerene nanoparticles (FNPs) observed at an elevated temperature in UHV using a needle sensor probe.[43] The needle sensor AFM probe is based on a quartz resonator with a sharp tungsten tip top, which operates at its resonance frequency of ~1 MHz. FNPs and FNWs were dispersed on a Si(111) substrate at room temperature (RT). The substrate was cleaned in UHV before the dispersion so that the atomic steps and terraces are visible. The FNPs of ~30 nm diameter were dispersed. Those FNPs were found to be decomposed and shrink at ~300°C. The decomposed C_{60} molecules are found to form a domain of a few nanometers high.

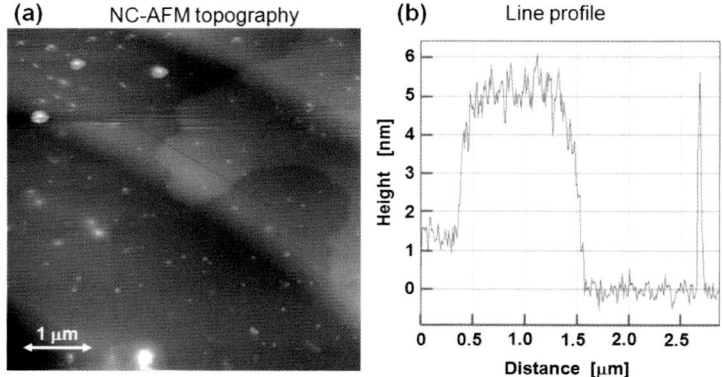

Figure 12.9 (a) An NCAFM image of FNPs dispersed on a Si(111) surface at ~300°C in ultrahigh vacuum (~10^{-8} Pa). The image was observed by using the needle sensor VT-SPM. (b) A line profile along the line indicated in (a). The C_{60} molecules domain of a few nanometers high is formed due to the thermal decomposition of the FNPs.

Temperature-dependent morphology change of an FNW on a Si(111) substrate in UHV is shown in Fig.12.10. The FNW of ~140 nm diameter at RT was found to shrink in height gradually down to ~15 nm with the temperature increase of up to ~600°C. Miyazawa et al. have already reported that the fullerene shell tubes can be formed by heating FNWs at 600–700°C in vacuum through the sublimation of inner C_{60} molecules.[44] The observed gradual shrinkage of FNW at the elevated temperatures might be caused by the diffusion and desorption process of the inner C_{60} molecules filled in the whisker. The outer shell is composed of amorphous carbon, which is developed from the original surface layer. It has been reported that illumination of visible or UV light on the surface of C_{60} crystals leads to polymerization of the surface layers.[45,46] Sakuma et al. have reported that the photopolymerized surface exhibits a high resistance to thermal sublimation so that the skins remained after sublimation of the inner part of the crystals.[47]

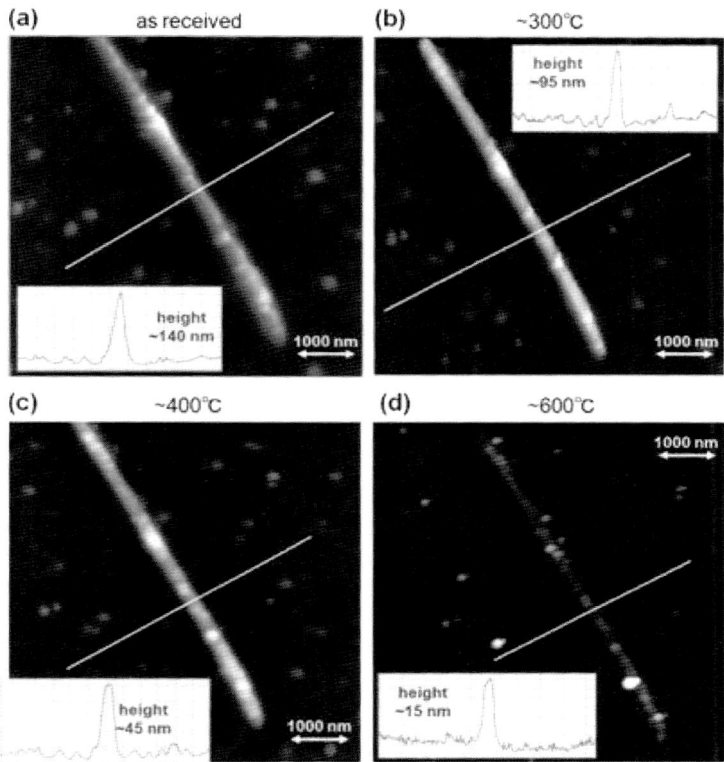

Figure 12.10 Temperature-dependent morphology change of an FNW dispersed on a Si(111) substrate in UHV. NCAFM images from (a) to (d) were measured by using a needle sensor VT-SPM, showing the shrinkage of the FNW at the elevated temperatures ranging from RT to ~600°C.

12.8 CONCLUSIONS AND OUTLOOK

This chapter has focused almost exclusively on the use of scanning probe microscopy for the in-depth characterization of the surfaces of FNMs. We have shown here that the rapid development of the surface nanocharacterization methodology has enabled us to clarify the basic property of novel FNMs. Nowadays, AFM has become one of the most powerful nanocharacterization tool. Static mode AFM can clarify the friction property on the surface of FNMs as well as the surface morphology. Dynamic mode AFM operated in the noncontact or intermittent contact mode can give us the mechanical and chemical information on the surface using the phase imaging. Other mechanical property analyses using SPM such as nanoindentation is also important and shall be applied in the near future.

Raw topography images of AFM generally suffer from various artifacts. The most common artifact is caused by the probe tip itself. Quantitative morphology analysis of the FNMs can be realized by the development of image reconstruction technique using probe shape functions.

One of the useful functions of FNMs is based on its n-type semiconducting property. Conductive AFM is a powerful tool to investigate the electric conductivity of FNMs at the nanoscale. A combination of conductive AFM and SAM has clarified that the intercalation of oxygen into the surface layer FNWs causes time-dependent insulating behavior.

In the end, the thermal stability of FNMs is introduced, which can be investigated by in situ microscopy measurements using the high-temperature SPMs in UHV.

Acknowledgments

This work was partially supported by the Japan Science and Technology Agency, NEDO International Joint Research Grant Program, and the World Premier International Research Center Initiative (WPI Initiative) on Materials Nanoarchitectonics, Ministry of Education, Culture, Sports, Science and Technology, Japan. We would like to thank those who provided unpublished materials that were used in this chapter.

References

1. K. Miyazawa, A. Obahashi, and M. Kuwabara, *J. Am. Ceram. Soc.*, **84**, 3037 (2001).
2. K. Miyazawa, Y. Kuwasaki, K. Hamamoto, S. Nagata, A. Obayashi, and M. Kuwabara, *Surf. Interf. Anal.*, **35**, 117 (2003).
3. D. Fujita, T. Kumakura, K. Onishi, and M. Harada, *Jpn. J. Appl. Phys.*, **42**, 1391 (2003).
4. M. Harada and D. Fujita, *J. Phys.: Conf. Ser.*, **159**, 012025 (2009).
5. G. Binnig and H. Rohrer, *Helv. Phys. Acta*, **55**, 726 (1982).
6. G. Binnig, C. F. Quate, and Ch. Gerber, *Phys. Rev. Lett.*, **56**, 930 (1986).
7. D. Fujita, T. Yakabe, H. Nejoh, T. Sato, and M. Iwatsuki, *Surf. Sci.*, **366**, 93 (1996).
8. J. T. Sadowski, R.Z. Bakhtizn, A.I. Oreshkin, T. Nishihara, A. Al-Mahboob, Y. Fujikawa, J. Nakajima, and T. Sakurai, *Surf. Sci.*, **601**, L136 (2007).
9. A. Marchenko and J. Cousty, *Surf. Sci.*, **513**, 233 (2007).
10. G. Meyer and N. Amer, *Appl. Phys. Lett.*, **53**, 1045 (1988).
11. D. Ruger, H. J. Mamin, and P. Güthner, *Appl. Phys. Lett.*, **55**, 2588 (1989).
12. N. Blanc, J. Brugger, N. F. de Rooj, and U. Dürig, *J. Vac. Sci. Technol. B*, **14**, 901 (1996).
13. R.C. Barret, M. Tortonese, and C. F. Quate, *Appl. Phys. Lett.*, **62**, 834 (1993).
14. G. Binnig, Ch. Gerber, E. Stoll, T. R. Albrecht, and C. F. Quate, *Europhys. Lett.*, **3**, 1281 (1987).
15. T. R. Albrecht and C. F. Quate, *J. Appl. Phys.*, **62**, 2599 (1987).
16. C. M. Mate, G. M. McClelland, R. Erlandsson, and S. Chiang, *Phys. Rev. Lett.*, **59**, 1942 (1987).
17. R. Garcia and R. Perez, *Surf. Sci. Rep.*, **47**, 197 (2002).
18. F. J. Giessibl, *Science*, **267**, 1451 (1995).
19. Q. Zhong, D. Inniss, K. Kjoller, and V. B. Elings, *Surf. Sci.*, **290**, L688 (1993).

20. S. N. Magonov, V. Elings, and M.-H. Whangbo, *Surf. Sci.*, **375**, L385 (1997).
21. D. Fujita, H. Itoh, S. Ichimura, and T. Kurosawa, *Nanotechnology*, **18**, 084002 (2007).
22. D. Fujita, K. Onishi, and M. Xu, *J. Phys.: Conf. Ser.*, **159**, 012002 (2009).
23. K. Onishi and D. Fujita, *J. Vac. Soc. Jpn.*, **51**, 165 (2008).
24. M. Xu, D. Fujita, and K. Onishi, *Rev. Sci. Instrum.*, **80**, 043703 (2009).
25. D. Keller, *Surf. Sci.*, **253**, 353 (1991).
26. H. Itoh, T. Fujimoto, and S. Ichimura, *Rev. Sci. Instrum.*, **77**, 103704 (2006).
27. M. Xu, D. Fujita, K. Onishi, and K. Miyazawa, *J. Nanosci. Nanotechnol.*, **9**, 6003 (2009).
28. T. L. Makarova, *Semiconductors*, **35**, 257 (2001).
29. R. Kitaura and H. Shinohara, *Jpn. J. Appl. Phys.*, **46**, 881 (2007).
30. K. Itaka, M. Yamashiro, J. Yamaguchi, M. Haemori, S. Yaginuma, Y. Matsumoto, M. Kondo, and H. Koinuma, *Adv. Mater.*, **18**, 1713 (2006).
31. H. Hoppe and N. S. Sariciftci, *J. Mater. Res.*, **19**, 1924 (2004).
32. K. Ogawa, T. Kato, A. Ikegami, H. Tsuji, N. Aoki, Y. Ochiai, and J. P. Bird, *Appl. Phys. Lett.*, **88**, 1121009 (2006).
33. Y. J. Xing, G. Y. Jing, J. Xu, D. P. Yu, H. B. Liu, and Y. L. Li, *Appl. Phys. Lett.*, **87**, 263117 (2005).
34. H.-X. Ji, J.-S. Hu, L.-J. Wan, Q.-X. Tang, and W.-P. Hu, *J. Mater. Chem.*, **18**, 328 (2008).
35. M. P. Larsson, J. Kjelstrup-Hansen, and S. Lucyszyn, *ECS Trans.*, **2**, 27 (2007).
36. M. Xu, Y. Pathak, D. Fujita, C. Ringor, and K. Miyazawa, *Nanotechnology*, **19**, 075712 (2008).
37. A. Hamed, Y. Y. Sun, Y. K. Tao, R. L. Meng, and P. H. Hor, *Phys. Rev. B*, **47**, 10873 (1993).
38. A. Tapponnier, I. Biaggio, and P. Günter, *Appl. Phys. Lett.*, **86**, 112114 (2005).
39. H. Habuchi, S. Nitta, D. Han, and S. Nonomura, *J. Appl. Phys.*, **87**, 8580 (2000).

40. K. Rauwerdink, J.-F. Liu, J. Kintigh, and G. P. Miller, *Microsc. Res. Tech.*, **70**, 513 (2007).

41. G. Radu, U. Memmert, and U. Hartmann, *Appl. Surf. Sci.*, **188**, 435 (2002).

42. J. Broekmaat, A. Brinkman, D. H. A. Blank, and G. Rijnders, *Appl. Phys. Lett.*, **92**, 043102 (2008).

43. W. Clauss, J. Zhang, D. J. Bergeron, and A. T. Johnson, *J. Vac. Sci. Technol. B*, **17**, 1309 (1999).

44. K. Miyazawa, J. Minato, H. Zhou, T. Taguchi, I. Homma, and T. Suga, *J. Eur. Ceram. Soc.*, **26**, 429 (2006).

45. M. Haluska, M. Haluska, H. Kuzmany, M. Vybornnov, P. Rogl, and P. Fejdi, *Appl. Phys. A*, **56**, 161 (1993).

46. J. Li, M. Ozawa, M. Kino, T. Yoshizawa, T. Mitsuki, H. Horiuchi, O. Tachikawa, K. Kishio, and K. Kitazawa, *Chem. Phys. Lett.*, **227**, 572 (1994).

47. H. Sakuma, M. Tachibana, H. Sugiura, K. Kojima, S. Ito, T. Sekiguchi, and Y. Achiba, *J. Mater. Res.*, **12**, 1545 (1997).

Chapter 13

STRUCTURAL AND THERMODYNAMIC PROPERTIES OF FULLERENE NANOWHISKERS

Hideaki Kitazawa and Kenjiro Hashi

National Institute for Materials Science, Tsukuba, Ibaraki 305-0047, Japan
Klitazawa.Hideaki@nims.go.jp

The motion of C_{60} molecules in fullerene nanowhiskers (C_{60}NWs) has been investigated by means of low-temperature X-ray diffraction, magnetic susceptibility, specific heat, and ^{13}C NMR (nuclear magnetic resonance). The temperature dependence of the lattice constant for C_{60}NWs is in good agreement with that for the pristine C_{60} powder, which demonstrates a large discontinuity at $T_c \sim 265$ K caused by the structural phase transition. However, the temperature dependence of magnetic susceptibility for C_{60}NWs exhibits a faint anomaly at T_c. The temperature dependence of specific heat for C_{60}NWs shows two small anomalies at 232 and 254 K. High-resolution ^{13}C NMR measurements of the C_{60}NWs indicate that both toluene and 2-propanol molecules interact with the C_{60}NW molecules in the suspension and the precipitate states obtained by using the liquid–liquid interfacial precipitation method. The broader linewidth of ^{13}C NMR in the dried C_{60}NWs compared with that in the pristine C_{60} powder indicates slower molecular

---

Fullerene Nanowhiskers
Edited by Kun'ichi Miyazawa
Copyright © 2012 Pan Stanford Publishing Pte. Ltd.
www.panstanford.com

reorientation in the dried C_{60}NWs compared with that in the pristine C_{60} powder. Since a rapid rotation of C_{60} molecules could be directly detected by ^{13}C NMR, polymer formation between C_{60} molecules is unlikely in C_{60}NWs at room temperature.

13.1 INTRODUCTION

Fullerene C_{60} nanowhiskers (C_{60}NWs) obtained by using the liquid–liquid interfacial precipitation (LLIP) method are less than 0.5 µm in diameter and greater than 100 µm in length.[1] The initial transmission electron microscopy (TEM) experiments suggested the existence of polymerization between C_{60} molecules along the growth axis.[1] However, subsequent X-ray diffraction (XRD) measurements revealed that the first precipitated crystals from the liquid phase have solvated structures with a hexagonal structure.[2] After the evaporation of solvent molecules in air, the structure rapidly changed into a face-centered cubic (fcc) structure such as a C_{60} solid. Since the Raman spectroscopic study revealed the photochemical polymerization,[3] the irradiation of the electron beam may affect the change in the interaction between C_{60} molecules. Thus, there is still uncertainty for the crystal structure of C_{60}NWs under various experimental conditions. It is well known that the solid C_{60} undergoes an orientational ordering transition at $T_c \sim 260$ K. The phase transition is from nearly free rotating state of the molecules forming the fcc structure (space group $Fm3m$) to a phase wherein the molecules undergo only a ratchet type of correlated rotation about <111> directions of a simple cubic (sc) structure (space group $Pa3$).[4,5] If the polymerization between C_{60} molecules occurs in C_{60}NWs, some dynamics such as a rotation of C_{60} molecules observed above T_c and structural phase transition at T_c should be affected in C_{60}NWs. Since the ordering transition has been investigated with a wide range of techniques such as X-ray,[4] neutron diffraction,[6] nuclear magnetic resonance (NMR),[5] and specific heat [7], it is worth understanding the dynamics of C_{60} molecules in C_{60}NWs by means of several kinds of techniques. In this chapter, we will demonstrate the structural and dynamical properties of C_{60}NWs from our recent measurements of low-temperature XRD, magnetic susceptibility, specific heat, and ^{13}C NMR for C_{60}NWs.[8]

13.2 EXPERIMENTAL PROCEDURES

First, the structural phase transitions of C_{60}NWs were investigated by the low-temperature X-ray diffraction, magnetic susceptibility, and specific heat measurements. The C_{60}NWs (dried C_{60}NWs in air) for above measurements were prepared by the LLIP method from a toluene solution of the pristine C_{60} powder (99.5%, MTR Ltd.) and isopropyl alcohol.[1] In order to determine the temperature dependence of lattice constants, the powder XRD experiments were performed by using a Rigaku diffractometer (CuKα radiation) with a low-temperature option. The powder samples were mixed with a standard Si material (NIST SRM 640b) and Apiezon-N grease to obtain the more precise lattice constants and make a good thermal contact on the Cu holder. All the peaks below $2\theta = 35°$ were well indexed by the fcc crystal symmetry of the solid C_{60}, except for small unknown peaks that may come from the remaining phases with a solvated structure or other foreign phases. Each lattice constant a of C_{60}NWs and pristine C_{60} powder was estimated from some assigned peaks below $2\theta = 35°$. The magnetic susceptibility $\chi(T)$ was measured in a magnetic field of 0.1 and 3 T for temperatures from 1.9 to 300 K by using a SQUID magnetometer (Quantum Design Co., Ltd.). The powder sample for the magnetic susceptibility measurement was wrapped in the Al(4N) sheet. The specific heat measurements were performed between 4 K and room temperature for pressed C_{60}NWs and pristine C_{60} powder with the relaxation method by a PPMS instrument (Quantum Design Co., Ltd).

Second, NMR measurements were performed at 11.7 T (125.8 MHz for ^{13}C) using a JEOL ECA500 spectrometer in order to investigate the interaction between solvent and C_{60}NW. Three C_{60}NW samples were prepared for NMR measurements: suspension C_{60}NW, precipitate C_{60}NW, and dried C_{60}NW. The suspension C_{60}NW was obtained from 2-propanol (isopropyl alcohol) and toluene solution of pristine C_{60} by using the LLIP method. Thus, the C_{60}NW particles exist in the solvent of 2-propanol and toluene in the suspension

C_{60}NW. The precipitate C_{60}NW was obtained from the suspension C_{60} NW by centrifugation. The dried C_{60}NW was obtained by drying the precipitate C_{60}NW in air at room temperature for more than 30 days. Both the suspension and the precipitate C_{60}NWs contain toluene and 2-propanol as solvents used in the LLIP method, while the dried C_{60}NW contains no solvent. A 4 mm sample tube was used for a magic-angle-spinning (MAS) NMR measurement. A Fourier-transform NMR spectrum was obtained from a free-induction-decay signal following a $\pi/2$ pulse of 2.4 µs.

13.3 STRUCTURAL PHASE TRANSITIONS OF C_{60}NWs

13.3.1 Low-Temperature X-Ray Diffraction

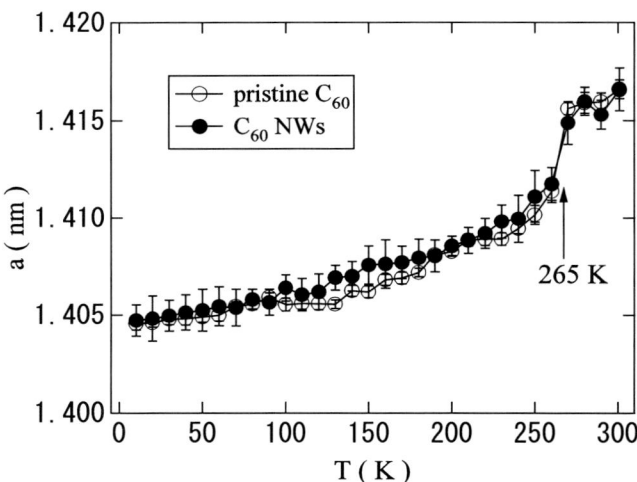

Figure 13.1 Temperature dependence of the lattice constant a for C_{60}NWs and pristine C_{60} powder. (Reprinted with permission from Kitazawa et al.,[8] © 2009, IOP Publishing Ltd.)

Figure 13.1 shows the temperature dependence of lattice constants for C_{60}NWs and pristine C_{60} powder. Both curves demonstrate large discontinuities at ~265 K, which correspond to the structural phase transition at $T_c \sim 260$ K from the sc structure at low temperatures to the fcc structure at high temperatures.[4]

Both curves are almost the same within the present experimental error bar. Overall temperature evolutions of lattice constants for both samples are in good agreement with the result determined from neutron diffraction,[6] except for the anomaly at $T_{sg} \sim 90$ K, which corresponds to the frozen rotation motion.

13.3.2 Magnetic Susceptibility

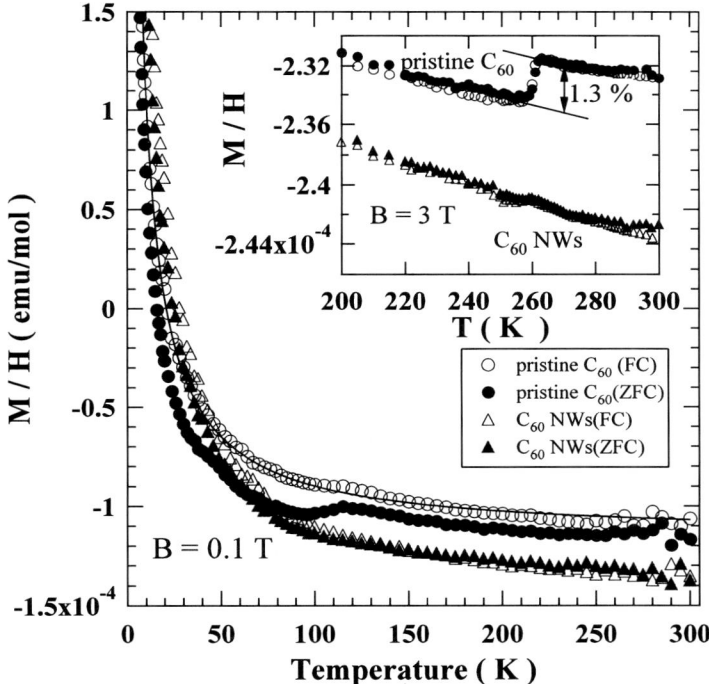

Figure 13.2 Temperature dependence of the magnetic susceptibility for pressed C_{60}NWs and pristine C_{60} powder. The solid lines represent the fitted curves by the Curie–Weiss law. The inset shows the precise data near 260 K. (Reprinted with permission from Kitazawa et al.,[8] © 2009, IOP Publishing Ltd.)

Figure 13.2 shows the temperature dependence of the magnetic susceptibility $\chi(T)$ for C_{60}NWs and pristine C_{60} powder. A clear steplike transition at 260 K was observed in the pristine C_{60}

powder, as shown in the inset of Fig. 13.2. This susceptibility discontinuity of ~1.3% is attributed to the change in the intramolecular geometry at an orientational order–disorder transition.[9] On the other hand, the discontinuity is very tiny but visible in C_{60}NWs. This reduction of anomaly at 260 K suggests that the structural phase transition is incomplete or short range in the C_{60}NWs below T_c. The overall field-cooled $\chi(T)$ curves of the C_{60}NWs and pristine C_{60} powder can be well fitted by the Curie–Weiss law $M/H = \chi(T) = C/(T - \theta_p) + \chi_0$, as shown by the solid curves in Fig. 13.2. In this equation, C, θ_p, and χ_0 are the Curie constant ($C = Np_{eff}^2/3k_B$, where N, p_{eff}, and k_B are the number of magnetic moments, the effective moment, and the Boltzmann constant, respectively), the paramagnetic Curie temperature, and the temperature-independent term, respectively. If the spin amplitude of magnetic moment is assumed to be 1/2, the number of magnetic moments corresponds to 1.20 and 0.693% in a unit cell for C_{60}NWs and pristine C_{60} powder, respectively. The estimated magnetic moments may originate from any magnetic impurity. The obtained parameters are listed in Table 13.1. The $\chi(T)$ curves of pristine C_{60} powder below ~100 K depend on zero-field-cooled (ZFC) and field-cooled (FC) processes, but the $\chi(T)$ curve of C_{60}NWs does not exhibit such a hysteresis. The hysteresis can be explained in terms of a transition into a frozen magnetic glass state below T_{sg} ~ 90 K.[10]

Table 13.1 Magnetic parameters of C_{60}NWs and pristine C_{60} powder

	χ_0 (emu/mol)	C [emu/ (mol K)]	θ_p (K)	N (/mol)	N (%/ unit cell)
C_{60}NWs	$-1.521(6) \times 10^{-3}$	$4.51(2) \times 10^{-3}$	$-1.61(2)$	7.24×10^{21}	1.20
Pristine C_{60}	$-1.154(4) \times 10^{-3}$	$2.60(2) \times 10^{-3}$	$-1.61(2)$	4.17×10^{21}	0.693

13.3.3 Specific Heat

Figure 13.3 shows the temperature dependence of specific heat for pristine C_{60} powder and pressed C_{60}NWs. A large peak at 254 K in the pressed pristine C_{60} powder must be caused by this structural phase transition.[7] The single anomaly observed in the pressed pristine C_{60}

splits into two small anomalies in the pressed C_{60}NWs at 254 K and 232 K. The former anomaly at 254 K is smaller than that for pressed pristine C_{60} powder. The latter anomaly at 232 K is not observed in the pressed pristine C_{60} powder. This splitting of anomaly at 254 K in the pressed pristine C_{60} powder is a strong evidence of suppression of the structural transitions in C_{60}NWs. The specific heat in pressed C_{60}NWs starts to deviate from 6 K to higher temperature. This means that the entropy in pressed C_{60}NWs is reduced up to room temperature as compared with that in pressed pristine C_{60} powder.

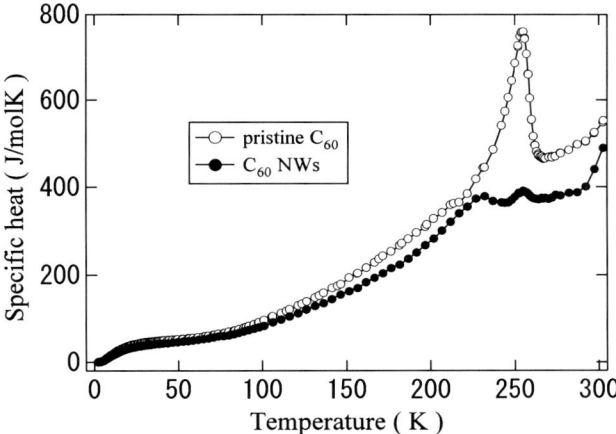

Figure 13.3 Temperature dependence of the specific heat for pressed C_{60} NWs and pristine C_{60} powder. (Reprinted with permission from Kitazawa et al.,[8] © 2009, IOP Publishing Ltd.)

13.4 STRUCTURAL PROPERTIES OF C_{60}NWs BY MEANS OF NMR

The NMR is one of the most powerful analytical tools to investigate structures of molecules from a microscopic point of view. Solution ^{13}C NMR measurements on C_{60} gave a clear evidence that a C_{60} molecule consists of exactly equivalent 60 carbon atoms and has a icosahedral symmetry.[11,12] NMR measurements on a solid-state C_{60} revealed that C_{60} molecules rotate rapidly and isotropically at room temperature, while C_{60} molecules jump between symmetry-equivalent orientations at low temperatures.[6,13] There are many other NMR studies of C_{60} and fulleride superconductors as reviewed in Ref. 14. On the other hand, structural study on C_{60}NWs from a microscopic point of view is limited. In this section, we present our recent NMR study on C_{60}NW.[8]

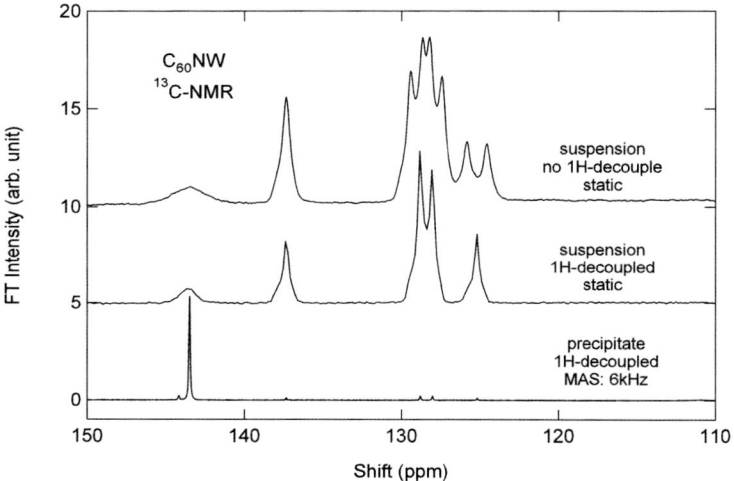

Figure 13.4 NMR spectra for the suspension C_{60}NW with and without proton (^{1}H) decoupling and for the precipitate C_{60}NW with a proton decoupling and MAS of 6 kHz. (Reprinted with permission from Kitazawa et al.,[8] © 2009, IOP Publishing Ltd.)

The NMR spectrum reflects local environment at a nucleus and dynamical property of a molecule. Figure 13.4 shows the NMR spectra for the suspension C_{60}NW with and without proton (^{1}H) decoupling. A proton-decoupled MAS NMR spectrum for the precipitate C_{60}NW is also shown in Fig. 13.4. The proton decoupling was performed by the WALTZ pulse sequence. A peak at 143 ppm is assigned to a signal from the C_{60} molecule, indicating that a carbon atom in a C_{60}NW has the sp^2 character as in a pristine C_{60}. The rests of the peaks are assigned to peaks from toluene of the solvent. The proton decoupling procedure eliminates the J coupling splits of the toluene peaks. It should be noted here that the peak of C_{60}NW is also narrowed by the proton decoupling. It seems that the origin of the linewidth of C_{60}NW is the dipole fields of the protons in solvent, because the NMR spectrum of C_{60}NW decoupled from the dipole fields by MAS shows a narrow single peak. This indicates that the C_{60}NW interacts with the proton of the solvent.

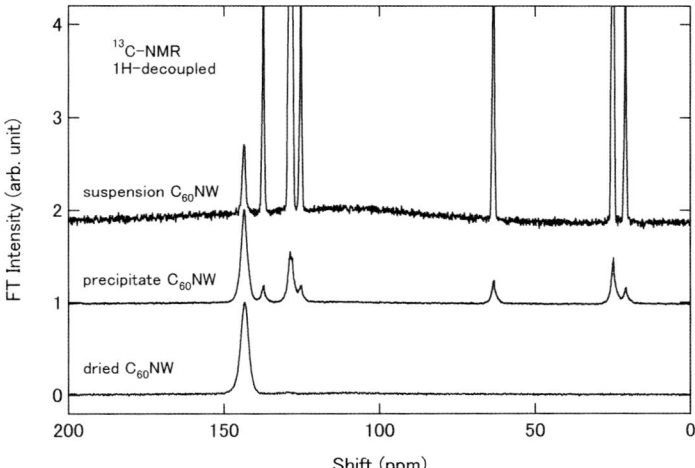

Figure 13.5 Proton-decoupled NMR spectra for suspension C_{60}NW, precipitate C_{60}NW, and dried C_{60}NW. (Reprinted with permission from Kitazawa et al.,[8] © 2009, IOP Publishing Ltd.)

In order to investigate the dynamical properties of the C_{60}NWs and solvent molecules, NMR measurements were performed for suspension C_{60}NWs, precipitate C_{60}NWs, and dried C_{60}NWs. Figure 13.5 shows the proton-decoupled NMR spectra for suspension C_{60}NWs, precipitate C_{60}NWs, and dried C_{60}NWs. It should be noted here that all peaks in the spectrum of precipitate C_{60}NWs are broader than those of suspension C_{60}NWs. This indicates that both C_{60}NWs and solvent molecules cannot reorient freely in the precipitate sample. The broader linewidth of the peaks for the solvent indicates that both the toluene and 2-propanol molecules are bound to C_{60}NWs in the precipitate sample. The linewidth of the peak for C_{60}NW in the precipitate sample is almost the same with that of the dried C_{60}NW, which is broader than that of pristine C_{60}. It seems that the broader linewidth originates from a recovery of the chemical shift anisotropy with a slower molecular reorientation.

The stronger bonding between C_{60} molecules may make the molecular reorientation slow. No peak, however, was observed for the polymerized C_{60} with the sp^3 character in this study. The stronger bonding between C_{60} molecules in C_{60}NWs is also expected from a shorter intermolecular distance of the C_{60}NWs along the growth axis as compared with the pristine C_{60} crystals.[1] Further studies are needed to clarify the nature of the stronger bonding.

In summary, from the NMR measurements for suspension, precipitate, and dried C_{60}NWs, it is confirmed that carbon atoms in a C_{60}NW obtained by the LLIP method have the sp^2 character as in a pristine C_{60}. C_{60} molecules in a C_{60}NW interact with both toluene and 2-propanol of the solvent. Molecular reorientation in C_{60}NW is expected to be slower than that in pristine C_{60}.

Acknowledgments

One of the authors, K.H., appreciates Dr. M. Murakami and Dr. K. Deguchi for technical support and fruitful discussions.

References

1. K. Miyazawa, Y. Kuwasaki, A. Obayashi, and M. Kuwabara, *J. Mater. Res.*, **17**, 83 (2002).
2. J. Minato and K. Miyazawa, *Carbon*, **43**, 2837 (2005).
3. M. Tachibana, K. Kobaysashi, T. Uchida, K. Kojima, M. Tanimura, and K. Miyazawa, *Chem. Phys. Lett.*, **374**, 279 (2003).
4. P. A. Heiney, J. E. Fischer, A. R. McGie, W. J. Pomanow, A. M. Denenstein, J. P. McCauley Jr, A.B. Smith III, and D. E. Cox, *Phys. Rev. Lett.*, **66**, 2911 (1991).
5. R. Tycko, G. Dabbagh, R. M. Fleming, R. C. Haddon, A. V. Makhija, and S. M. Zahurak, *Phys. Rev. Lett.*, **67**, 1886 (1991).
6. W. I. F. David, R. M. Ibberson, T. J. S. Dennis, J. P. Hare, and K. Parssides, *Europhys. Lett.*, **18**, 219 (1992).
7. T. Matsuo, H. Suga, W. I. F. David, R. M. Ibberson, P. Bernier, A. Zahab, C. Fabre, A. Rassat, and A. Dworkin, *Solid State Commun.*, **83**, 711 (1992).
8. H. Kitazawa, K. Hashi, T. Wuernisha, K. Hotta, C. L. Ringor,

T. Furubayashi, A. Goto, T. Shimizu, and K. Miyazawa, *J. Phys.: Conf. Ser.*, **159**, 012022 (2009).

9. W. Luo, H. Wang, R. S. Ruoff, J. Cioslowski, and S. Phelps, *Phys. Rev. Lett.*, **73**, 186 (1994).

10. V. Buntar, H. W. Weber, and M. Ricco, *Solid State Commun.*, **98**, 175 (1995).

11. R. Taylor, J. P. Hare, A. K. Abdul-Sada, and H. W. Kroto, *J. Chem. Soc. Chem. Commun.*, 1423 (1990).

12. R. D. Johnson, G. Meijer, and D. S. Bethune, *J. Am. Chem. Soc.*, **112**, 8983 (1990).

13. C. S. Yannoni, R. D. Johnson, G. Meijer, D. S. Bethune, and J. R. Salem, *J. Phys. Chem.*, **95**, 9 (1991).

14. C. H. Pennington and V. A. Stenger, *Rev. Mod. Phys.*, **68**, 855 (1996).

Chapter 14

HIGH-TEMPERATURE HEAT TREATMENT OF FULLERENE NANOFIBERS

Ryoei Kato,[a] Kun'ichi Miyazawa,[a] Toshiyuki Nishimura,[a] Zheng-ming Wang,[b] and Tokushi Kizuka[c]

[a] *National Institute for Materials Science, 1-1, Namiki, Tsukuba, Ibaraki 305-0044, Japan*

[b] *Energy Technology Research Institute, National Institute of Advanced Industrial Science and Technology (AIST), 16-1 Onogawa, Tsukuba, Ibaraki 305-5869, Japan*

[c] *Institute of Materials Science, University of Tsukuba, 1-1-1, Tennoudai, Tsukuba, Ibaraki 305-8573, Japan*

kato.ryoei@nims.go.jp

The fullerene nanofibers composed of C_{60}, "C_{60} nanofibers," were treated at various temperatures up to 3000°C in vacuum or in an Ar atmosphere. The structural characteristics of the heat-treated C_{60} nanofibers were examined by transmission electron microscopy and Raman spectroscopy. By high-temperature heat treatment, C_{60} nanofibers turned to the glassy carbon composed of randomly oriented graphitic ribbons. The interlayer spacing of the graphitic ribbons was found to decrease with the increasing number of the stacked layers; it became the (0002) spacing of graphite when the number was 53.

Fullerene Nanowhiskers
Edited by Kun'ichi Miyazawa
Copyright © 2012 Pan Stanford Publishing Pte. Ltd.
www.panstanford.com

14.1 INTRODUCTION

As reported in the introductory chapter, fine needlelike crystals composed of fullerenes, "fullerene nanowhiskers," can be synthesized by using the liquid–liquid interfacial precipitation method (the LLIP method). The fullerene nanowhiskers shown here have nontubular structures. On the other hand, the tubular fullerene nanowhiskers are especially called fullerene nanotubes and can also be synthesized by using the LLIP method.

The C_{60} nanofibers, which include both tubular and nontubular fullerene nanowhiskers, have diameters less than 1000 nm, and the dried fullerene nanofibers have the face-centered cubic (fcc) structure that is identical with the fcc structure of pristine C_{60} crystal.[1-6] The electrical resistivity of the C_{60} whiskers with diameters greater than 650 nm is higher than 3 Ω cm.[7] However, the C_{60} whiskers acquire a high electrical conductivity when they are heated at high temperature, resulting in the structural transition from van der Waals crystals to amorphous carbon.[8,9,10] In the present chapter, we describe the heat-treatment process of the C_{60} nanofibers at various temperatures up to 3000°C. The structural characteristics of the heat-treated samples will be shown by optical microscopy, scanning electron microscopy, transmission electron microscopy (TEM), and Raman spectroscopy.

14.2 HEAT TREATMENT OF C_{60} NANOFIBERS

Figure 14.1 C_{60} nanofibers vacuum-encapsulated in a quartz tube (<10^{-5} Pa) in order to suppress the sublimation of C_{60} upon heat treatment at 900°C.

The C_{60} nanofibers were prepared by using the LLIP method. The grown C_{60} nanofibers in a glass bottle were collected by a centrifugal separation method or filtration. The samples were vacuum-encapsulated in silica tubes to suppress the sublimation of C_{60}, as shown in Fig. 14.1.

The C_{60} nanofibers in the silica tubes were heated for 1 h at 900°C using a muffle furnace and collected from the silica tubes. These heat-treated C_{60} nanofibers were further heat-treated at 2000, 2500, and 3000°C for 1 h in an Ar atmosphere using an ultrahigh temperature furnace and graphite crucibles.

The structure of samples was examined by the use of high-resolution transmission electron microscopes (JEM-4000, JEM-2010, and JEM-2000, JEOL) and a microscopic Raman spectrophotometer (NRS-3100, JASCO).

14.3 TEM OBSERVATION OF THE HEAT-TREATED C_{60} NANOFIBERS

Figures 14.2 and 14.3 show both the low-magnification and high-resolution TEM images and the selected-area electron diffraction patterns (SAEDPs) of the as-grown and heat-treated C_{60} nanotubes. As shown in the SAEDP of Fig. 14.2b, the as-grown C_{60} nanotubes have a single crystalline structure. However, the C_{60} nanotubes heated at 900°C turned to an amorphous structure, as shown in Fig. 14.3b,c.

Figure 14.4 shows Raman spectra of the heat-treated C_{60} nanowhiskers, the as-grown C_{60} nanowhiskers, and graphite. The Raman spectrum of as-grown C_{60} nanowhiskers dried in air at room temperature is similar to that of pristine C_{60}.[11] Hence, the C_{60} nanowhiskers without the heat treatment are found to be composed of the C_{60} molecules that are loosely bound via van der Waals forces, similar to the case of pristine fcc crystals of C_{60}. The peaks observed for the as-grown C_{60} nanowhiskers are not observed in the C_{60} nanowhiskers heated at 900°C. This result shows that the spherical C_{60} cages were destroyed by the high-temperature heat treatment. It is also found that the Raman spectrum of the C_{60} nanowhiskers heated at 3000°C is similar to that of glassy carbon.[12,13]

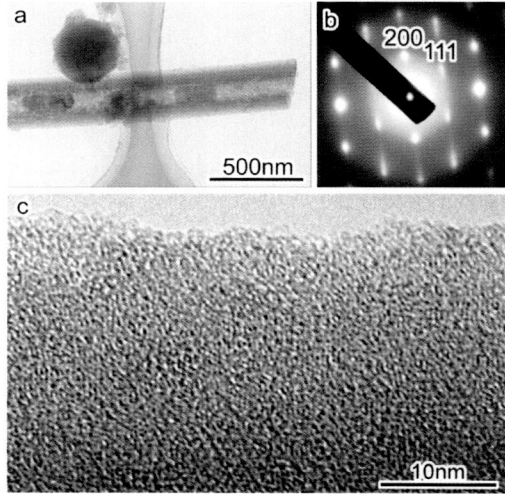

Figure 14.2 (a) Bright-field TEM image, (b) SAEDP, and (c) high-resolution TEM image of an as-grown C_{60} nanotube. (Reprinted with permission from Kato et al.,[10] © 2009, IOP Publishing Ltd.)

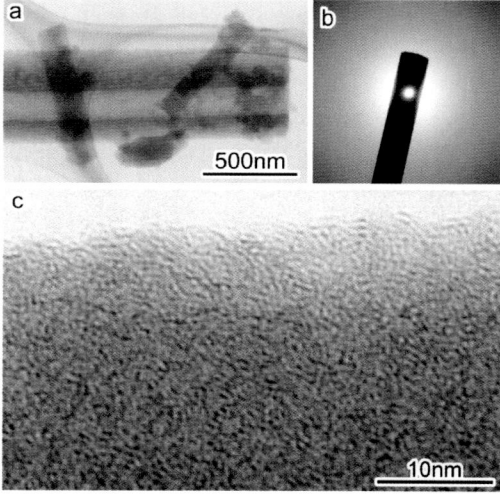

Figure 14.3 (a) Bright-field TEM image, (b) SAEDP, and (c) high-resolution TEM image of a C_{60} nanotube heated at 900°C. (Reprinted with permission from Kato et al.,[10] © 2009, IOP Publishing Ltd.)

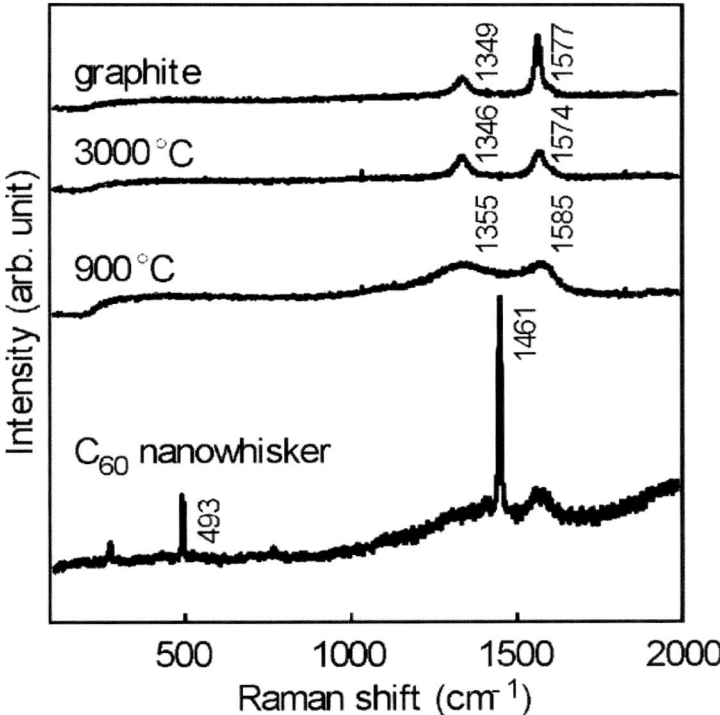

Figure 14.4 Raman spectra of the as-grown C_{60} nanowhiskers, the C_{60} nanowhiskers heat-treated at 900 and 3000°C, and graphite obtained by using a microscopic Raman spectrophotometer with a laser excitation wavelength of 532 nm. (Reprinted with permission from Kato et al.,[10] © 2009, IOP Publishing Ltd.)

Figure 14.5 shows TEM images of the C_{60} nanowhiskers which were heated at 2000, 2500, and 3000°C. By the high-temperature heat treatment, the C_{60} nanowhiskers turned to the glassy carbon nanofibers that are composed of randomly oriented graphite ribbons. This is consistent with the above result of Raman spectrometry. Figure 14.6 shows a bright-field TEM image of a C_{60} nanotube heated at 3000°C. It is found that the C_{60} nanotubes heated even at such a high temperature can retain their original tubular morphology.

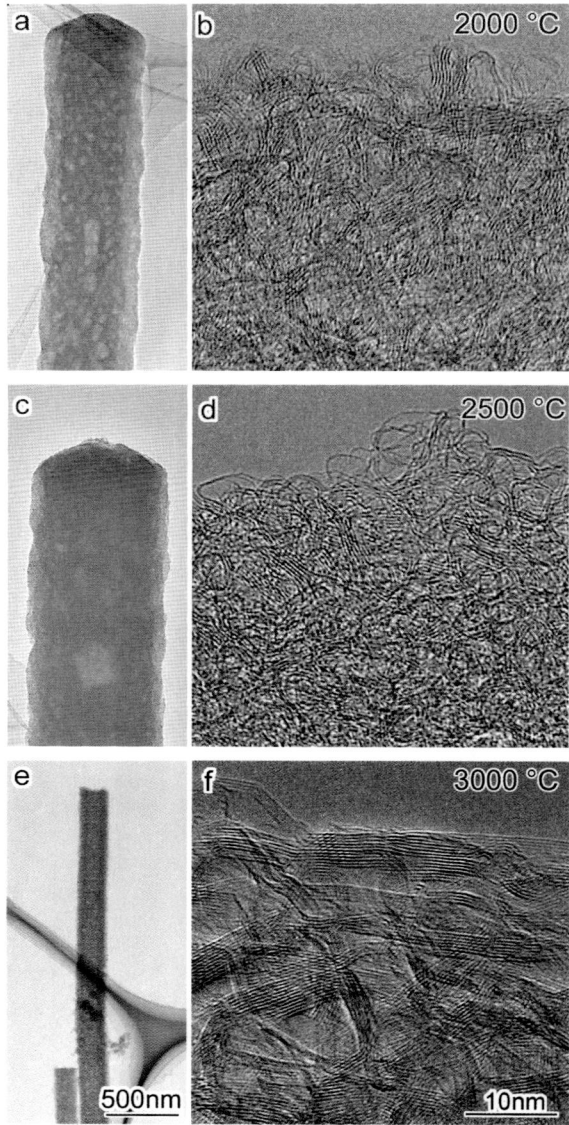

Figure 14.5 (a, c, e) Bright-field TEM images and (b, d, f) high-resolution TEM images of the C_{60} nanowhiskers heated at 2000, 2500, and 3000°C in Ar, respectively. (Reprinted with permission from Kato et al.,[10] © 2009, IOP Publishing Ltd.)

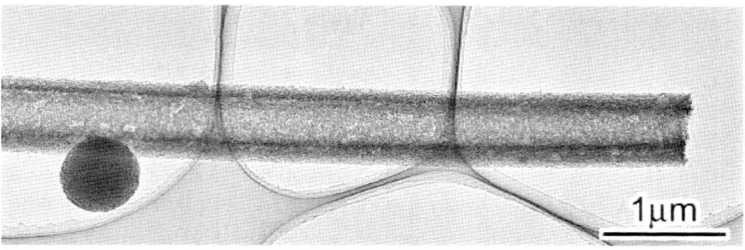

Figure 14.6 Bright-field TEM image of a C_{60} nanotube heated at 3000°C in Ar. (Reprinted with permission from Kato et al.,[10] © 2009, IOP Publishing Ltd.)

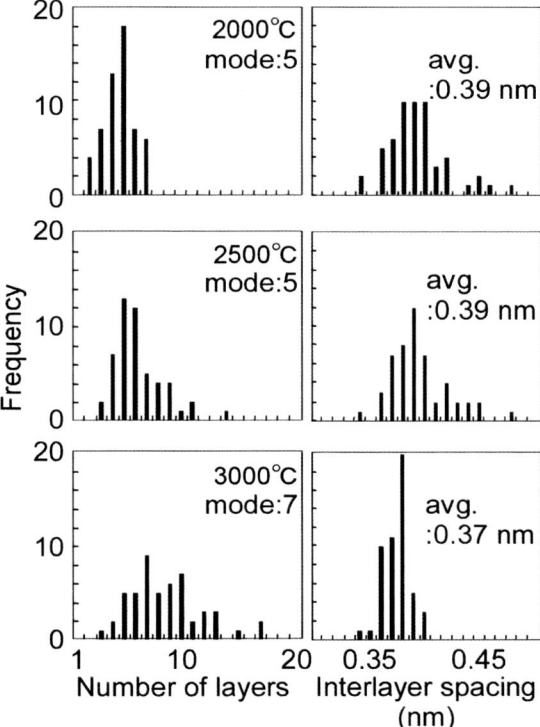

Figure 14.7 Histograms for the number of layers and the interlayer spacing of graphitic ribbons of the heat-treated C_{60} nanowhiskers. (Reprinted with permission from Kato et al.,[10] © 2009, IOP Publishing Ltd.)

The distribution of layer number and the interlayer spacing of the graphitic ribbons of the C_{60} nanowhiskers heated at each temperature were calculated, as shown in Fig. 14.7. The maximum layer number of the graphitic ribbons is 17.

With the increasing temperature of heat treatment, the number of layers increased whereas the average interlayer spacing decreased. Figure 14.8 shows the relationship between the number of layers (x) and the average interlayer spacing (y) of the graphitic ribbons observed in Fig. 7, where a linear relationship expressed by $y = -0.001x + 0.388$ is derived. It is assumed that the interlayer spacing of the heat-treated C_{60} nanofibers with 53 layers becomes 0.3354 nm of the (0002) spacing of graphite.[13]

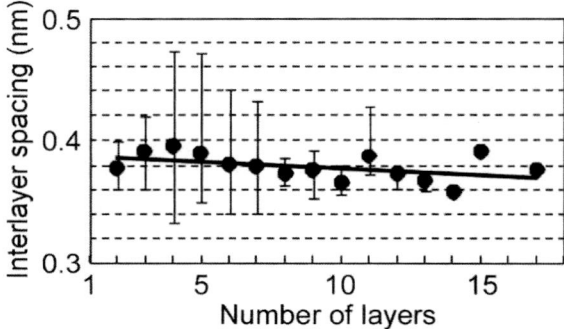

Figure 14.8 Relationship between the number of graphitic layers and the average interlayer spacing of the heat-treated C_{60} nanowhiskers in Fig. 14.7. (Reprinted with permission from Kato et al.,[10] © 2009, IOP Publishing Ltd.)

The as-prepared C_{60} nanowhiskers have an electrical resistivity greater than 3 Ω cm.[7] However, the electrical conductivity of the C_{60} whiskers heat-treated at high temperature in vacuum was improved owing to their transition into amorphous or glassy carbons. For example, a C_{60} whisker of 2.2 μm diameter showed a resistivity of 0.037 Ω cm and a C_{60} whisker of 2.9 μm diameter showed a resistivity of 0.042 Ω cm after the heat treatment at 1100°C for 30 min.[8]

The specific surface area of the C_{60} nanotubes heated at 200°C for 1 h in vacuum was 26.2 m^2/g.[4] However, the C_{60} nanofibers became porous by heating at higher temperatures and acquired higher specific surface areas. For example, the C_{60} nanowhiskers heated at 900°C in vacuum showed a specific surface area of 195 m²/g from N_2 adsorption.[14] Contrary to this, the specific surface area decreased as the heat treatment temperature increased higher than 900°C, and the C_{60} nanowhiskers heated at 2500°C showed a smaller value, i.e., 80 m²/g, corresponding to the structural densification caused by the development of graphitic layers, as observed in Fig. 14.5.[14]

In our previous experiment, an amorphous carbon nanofiber prepared by heating C_{60} nanowhiskers in vacuum at 1100°C was further electrically heated with the Joule heating in a transmission electron microscope. As a result, spherical hollow nanocapsules with carbon multiwalls, "multiwall fullerenes," were formed, as shown in Fig. 14.9.[9] Mordkovich et al. also found that the multiwall fullerenes are formed by heating the laser pyrolysis carbon blacks at 3000°C.[15] However, in our present experiment, no hollow carbon capsules were observed even after heat treatment at 3000°C. These results show that the structure of heat-treated carbons depends on the individual heat treatment condition, which leads to the formation of various kinds of novel nanocarbons.

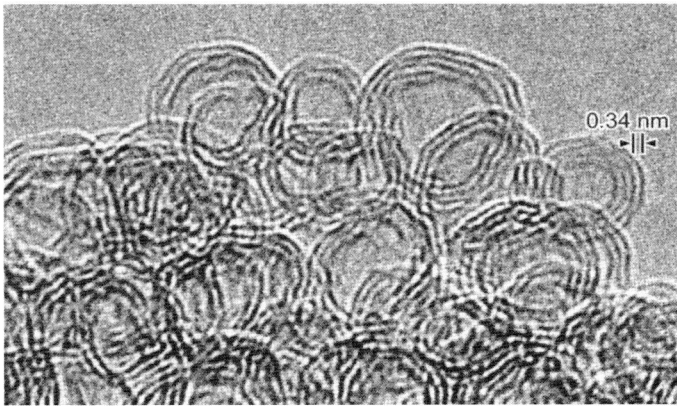

Figure 14.9 Multilayered carbon nanocapsules formed in a transmission electron microscope by the electrical resistive heating of an amorphous carbon nanofiber that was prepared by heating a C_{60} nanowhisker in vacuum. (Reprinted with permission from Asaka et al.,[9] © 2006, American Institute of Physics.)

The electrically conductive carbon nanofibers prepared by the heat treatment of C_{60} nanofibers at high temperatures contain no metal catalysts that are generally used in the chemical vapor deposition synthesis of carbon nanotubes, and hence they are expected to be utilized as new pure carbon electrode materials for the carrier of platinum catalysts used in fuel cells. In fact, the C_{60} nanowhiskers heated at 900°C in vacuum and loaded with platinum catalysts were successfully applied for a direct methanol fuel cell that was fabricated by the nanoimprint technology.[16]

14.4 CONCLUSIONS

The above research can be summarized as follows:
1. Both C_{60} nanowhiskers and nanotubes became amorphous by the heat treatment at 900°C in vacuum. These C_{60} nanofibers heated at temperatures higher than 2000°C exhibited the glassy carbon structures composed of randomly oriented graphitic ribbons.
2. The interlayer spacing of graphitic ribbons of the heat-treated C_{60} nanofibers was found to decrease with the increasing number of stacked layers; it became the (0002) plane spacing of graphite when the layer number is 53.
3. The C_{60} nanofibers became porous by the high-temperature heat treatment owing to the development of graphitic ribbons. The tubular morphology of the as-grown C_{60} nanotubes, however, could be maintained during the heat treatment even up to 3000°C.

Acknowledgments

The authors are grateful to Mr. H. Tsunakawa (The University of Tokyo) and Mr. K. Kurashima (National Institute for Materials Science) for the use of high-resolution transmission electron microscope.

References

1. K. Miyazawa, A. Obayashi, and M. Kuwabara, *J. Am. Ceram. Soc,.* **84**, 3037 (2001).
2. J. Minato and K. Miyazawa, *Diam. Relat. Mater.*, **15**, 1151 (2006).
3. K. Miyazawa, Y. Kuwasaki, K. Hamamoto, S. Nagata, A. Obayashi, and M. Kuwabara, *Surf. Interface Anal.*, **35**, 117 (2003).
4. C. Ringor and K. Miyazawa, *Diam. Relat. Mater.*, **17**, 529 (2008).
5. K. Miyazawa and C. Ringor, *Mater. Lett.*, **62**, 410 (2008).
6. K. Miyazawa, Y. Kuwasaki, A. Obayashi, and M. Kuwabara, *J. Mater. Res.*, **17**, 83 (2002).
7. M. P. Larsson, J. Kjelstrup-Hansen, and S. Lucyszyn, *ECS Trans.*, **2**, 27 (2007).
8. K. Miyazawa, J. Minato, H. Zhou, T. Taguchi, I. Honma, and T. Suga, *J. Eur. Ceram. Soc.*, **26**, 429 (2006).
9. K. Asaka, R. Kato, Y. Maezono, R. Yoshizaki, K. Miyazawa, and T. Kizuka, *Appl. Phys. Lett.*, **88**, 051914 (2006).
10. R. Kato, K. Miyazawa, T. Nishimura, and Z.-M. Wang, *J. Phys.: Conf. Ser.*, **159**, 012024 (2009).
11. M. C. Martin, D. Koller, A. Rosenbergand, C. Kendzior, and L. Mihaly, *Phys. Rev. B*, **51**, 3210 (1995).
12. S. K. Diane and B. W. William, *J. Mater. Res.*, **4**, 385 (1998).
13. J. Robertson, *Adv. Phys.*, **35**, 317 (1986).
14. Z.-M. Wang, R. Kato, K. Hotta, and K. Miyazawa, *J. Phys.: Conf. Ser.*, **159**, 012013 (2009).
15. V. Z. Mordkovich, A. G. Umnov, T. Inoshita, and M. Endo, *Carbon*, **37**, 1855 (1999).
16. Q. Wang, Y. Zhang, K. Miyazawa, R. Kato, K. Hotta, and T. Wakahara, *J. Phys.: Conf. Ser.*, **159**, 012023 (2009).

Chapter 15

ELECTRONICS DEVICE APPLICATION OF FULLERENE NANOWHISKERS

Yuichi Ochiai,[a] Nobuyuki Aoki,[a] and Jonathan Paul Bird[b]

[a] *Graduate School of Advanced Integration Science, Chiba University, 1-33 Yayoi-cho, Inage-ku, Chiba-city, Chiba 263-8522, Japan*
[b] *Department of Electrical Engineering, University at Buffalo, The State University of New York, Buffalo, NY 14260-1920, USA*
ochiai@faculty.chiba-u.jp

C_{60} fullerene nanowhiskers (FNWs) have been investigated for electronics device applications. Nanofabrication for C_{60} FNWs has been developed for device manufacturing based on simple synthesis and semiconductor microprocessing of C_{60}. Several kinds of disadvantages, including carrier suppression, were overcome under the control of the device structure. Electrical transport is explained and the device working characteristics of the field-effect transistor are also discussed.

Fullerene Nanowhiskers
Edited by Kun'ichi Miyazawa
Copyright © 2012 Pan Stanford Publishing Pte. Ltd.
www.panstanford.com

15.1 INTRODUCTION

There is much interest in nanostructured carbon materials for electronics device applications, since organic thin-film field-effect transistors (FETs) have been widely investigated from the perspective of developing near-future device applications.[1] Especially, the practical performance of C_{60} fullerene and carbon nanotubes should be tested in standard device components, such as FETs, as tentative device elements in practical circuits. The important criteria for the performance of such FETs are commonly considered to be a field-effect electron mobility of 1 cm^2/(V s) or more and an on/off ratio of the order of 10^6. In the case of the well-known organic semiconductor, a pentacene mobility of 1.5 cm^2/(V s) was achieved using one monolayer based on a SiO$_2$ surface treatment technique. However, such FETs show only p-type character in their operation. In C_{60} fullerenes, the thin film FET performance is n-type in character but the mobility is relatively low, while a higher mobility has been reported recently.[2] Therefore, improved performance of n-type C_{60} FETs is highly desired. An additional difficulty for this system is that the n-type behavior of C_{60} FETs is very sensitive to oxygen or ambient atmosphere, since the conductance quickly decreases by several orders due to oxygen adsorption in air. Consequently, usual device operation is achieved only for C_{60} FETs under vacuum. It is considered that the absorbed oxygen generates deep trap centers and suppresses carrier transport in C_{60} thin films. By covering the FET with a protective layer, such as a thin insulating layer of alumina, one can prevent carrier suppression in C_{60} thin films.[3] Especially, the alumina covering can also be used for a gate insulator with a large dielectric constant in order to obtain good electric breakdown properties; however, special handling and processing of alumina is needed.

Also there exist a few problems that arise in the practical application of C_{60} fullerenes and involve the use of nanofabrication processes based on electron-beam (E-B) lithography. On exposure of an E-B on C_{60}, its surface gets damaged and its resistance largely increases. Several preparation processes performed before and after E-B lithography clearly affect the electrical quality of C_{60} layers.

Therefore, there is a need to produce such nanoscale samples without using ordinary nanofabrication processes. One possible approach involves using nanosized C_{60} single crystals. A novel approach for synthesizing fibrous C_{60} crystals has been developed by using the method of liquid–liquid interface precipitation (LLIP). This yields C_{60} fullerene nanowhiskers (FNWs) with uniform nanoscale cross section over lengths that can easily exceed several microns.[4,5] In fact, several approaches to the solution-based growth of C_{60} crystals have been tried and reported recently due to the interest in using them as a nanotechnology candidate for smart starting materials in carbon-based nanoelectronics. Needlelike C_{60} structures have been obtained by slowly evaporating solutions of C_{60} in 1,2-dichloroethane (DECAN) at room temperature[6] or by toluene-containing C_{60}.[7] By using the LLIP method,[4,5] many fibrous C_{60} FNWs have been obtained. Those precipitated FNWs have been shown to exhibit single-crystal structure and have submicron diameters and lengths of greater than 100 µm. Furthermore, the FNWs have been confirmed that the fibers are mainly composed of C_{60} molecules as studied by Fourier transform infrared spectra.[8]

Unfortunately, however, the performance of C_{60}-based FETs is known to degrade on exposure to air, since adsorbed oxygen creates a deep trapping level that suppresses carrier transport.[3] Consequently, to date, it has only been possible to measure the electrical characteristics of FNW FETs under vacuum conditions after performing thermal annealing to remove oxygen and other adsorbate materials. One of the characteristics of C_{60} FNWs synthesized by using the LLIP method is that the as-grown, solvated C_{60} FNWs exhibit a hexagonal crystalline structure in contrast to the face-centered cubic (fcc) structure that is typical of C_{60} films at room temperature (RT). The hexagonal structure is converted to the normal fcc one, however, after evaporating the solvent from the crystal, and this change in the crystal structure is known to lead to the introduction of a high density of dislocations.[9] As a result, the C_{60} FNW FETs studied to date have typically suffered from pronounced crystalline disorder. In this chapter, however, we discuss an approach that allows FET operation of FNW FETs even in a nitrogen environment, without the need for vacuum isolation, thereby opening up the possibility of developing technological applications based on these devices.

C_{60} FNWs may eventually provide an important material in near-future nanotechnology. In spite of their potential importance for carbon-based nanoelectronics, however, there have been few reports of the electrical properties of these FNWs to date. Herein, we first discuss the basic electronic transport in C_{60} thin film (TF)-FET because almost similar transports have been observed as in C_{60} FNWs. Next, we discuss two attempts to fabricate FET structures incorporating either multiple[10] or single[11] C_{60} FNWs. In the last discussion, as for practical applications, it is desirable and also very important to be able to operate such FETs under ambient conditions.

15.2 ELECTRONIC PROPERTIES OF THIN FILM C_{60} FET

With regard to device applications of C_{60} fullerene, TF-FETs were first studied by Paloheimo[12] in 1992 and also fabricated by Haddon[13] in 1995. They found that C_{60} FETs worked as an n-type FET and that the electron mobility was as high as ~0.1 cm^2/(V s). However, their conductance shows a strong decrease with lowering temperature[12,14] due to a large amount of charge traps in the pseudogap of the electron states in a C_{60} thin film. Such trapped states likely originate in structural disorder, grain boundary, and/or polaron-type trappings introduced by guest impurities. A well-known oxygen adsorption effect in C_{60} most likely comes from the latter case and results in a drastic drop of the conductance. However, we can suppress the influence of this effect by passivity coverage[3] or vacuum heating.[15] Also, it is well known that the thermal annealing effect is effective for carrier conduction in terms of C_{60} crystalline growth.[16,17] Without annealing, an amorphous C_{60} thin film is obtained and the measurement shows a similar conductance decrease in the wide temperature range from 400 to 100 K. In fact, after annealing, the conductance becomes weakly dependent on temperature in comparison to the case without annealing. However, it is not so simple to explain the annealing effect with a charge transport mechanism in C_{60} so that we need to make further investigations not only of the barrier height at C_{60} grain boundary but also of the electron injection process into C_{60} at the interface at the source–drain electrodes or the gate dielectric. Moreover, a variable range hopping (VRH) has been observed in the conductance of C_{60} FET below 100 K. Conducting carriers in VRH can transport by hopping among electronic states at the Fermi level, which enables us to estimate the density of charge trapping states in the pseudogap of a C_{60} thin film. From the thermal annealing results of the VRH at various source–drain separations (L_{sd}) of C_{60} TF-FET, an efficiency of crystalline growth against L_{sd}[16,17] is very sensitive and important for the C_{60} device fabrication process.

15.3 ELECTRICAL PROPERTIES AND DEVICE FABRICATION OF C_{60} FNW

In order to fabricate C_{60} FNW-FETs, a clear analysis of their structural and electrical properties is important and needed. Basically, the transport properties of C_{60} FNW-FETs are similar to those of C_{60} TF-FETs, and so the basic transport properties of a C_{60} thin film is very important for realizing and analyzing C_{60} FNW-FETs. For example, thermally annealed FNWs show X-ray powder-diffraction spectra that are consistent with a similar fcc structure to that of C_{60} bulk crystals, although with a slightly reduced lattice constant ($a = 1.39$ nm). Also, the FNW-FETs exhibit n-channel enhancement-type transport behavior and their carrier mobility is estimated to be the order of 10^{-2} cm^2/(V s) under vacuum conditions at room temperature.[10] FETs based on thin films of organic molecules have recently been studied by many groups.[8,18] Among these, C_{60} TF-FETs have been found to show n-channel enhancement-type properties, with one of the higher mobilities [$\mu = 0.1$–0.5 cm^2/(V s)] reported for organic thin film n-channel FETs.[1,13,19] When fabricating such devices, it has been popular to use heating[16,20] or hydrophobic treatment[21] to enhance molecular accumulation and thereby obtain large organic crystals. In another approach, the size of the device is reduced to a scale comparable to that of the small organic crystals that are usually obtained by conventional crystal growth methods.[22] In C_{60} TF-FETs, a particular problem is that the thin film obtained by high-vacuum thermal deposition is typically polycrystalline and this granular nature prevents the achievement of devices with higher mobility.

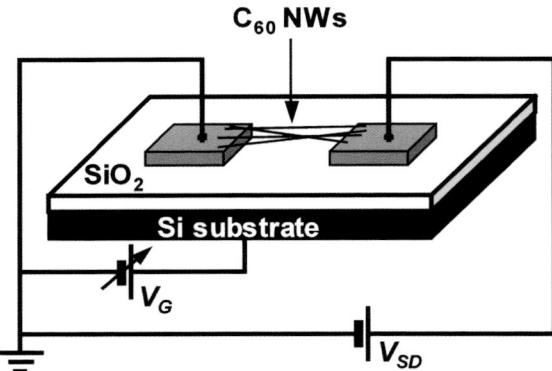

Figure 15.1 Schematic illustration of the structure of a C_{60} FNW-FET device. About 50 C_{60} FNWs bridge the electrodes in the device whose characteristics are reported here.

Here, it can be shown from the results of studies of C_{60} FETs that the semiconducting channels that are formed are connected in parallel by large numbers of single-crystal C_{60} FNWs. X-ray diffraction and atomic force microscope (AFM) inspection were used to characterize their structural properties. Our studies reveal that the structure of the FNWs is slightly changed by thermal annealing, after which a clear signature of an fcc structure is observed in the diffraction spectrum. Electrical properties of FETs have been investigated by incorporating large numbers of the FNWs as the channel of the device. These FETs show n-channel enhancement-type behavior, with a room temperature carrier mobility estimated to be 2×10^{-2} cm^2/(V s). The result demonstrates that C_{60} FNWs are an important potential candidate for high-mobility FETs based on organic materials.

The C_{60} FNWs were grown by means of the LLIP method using a system of C_{60}-saturated m-xylene and isopropyl alcohol. The C_{60}-saturated m-xylene solution was placed in a glass bottle, after which isopropyl alcohol was added so as to realize a liquid–liquid interface of saturated m-xylene solution and isopropyl alcohol. In order to promote the FNW growth, the bottle was capped and left undisturbed at 280 K for 7 days. The structural analysis of pristine and annealed (24 h at ~440 K and ~2 × 10^{-6} Torr) NWs was performed by X-ray powder diffraction using the CuKα line. A schematic diagram showing the structure of the C_{60}-FNW FETs that were fabricated in this study is shown in Fig. 15.1. A heavily doped (5 × 10^{-3} Ω cm) Si wafer was used as the gate substrate, with a thermally grown SiO_2 (950 nm) insulating top layer. Ti/Au (3/7 nm) source and drain electrodes were put on the surface of SiO_2 by photolithography and lift-off processes. The device had a channel length of 4 μm and about 50 C_{60} FNWs were used to bridge the source and drain electrodes by instillation, using a micropipette, of solution in the glass bottle containing C_{60} FNWs. In order to remove absorbed oxygen and solvents, the fabricated devices were annealed for about 24 h at ~440 K and ~10^{-6} Torr. Their FET characteristics were then measured in situ, without exposing the devices to the ambient atmosphere.

 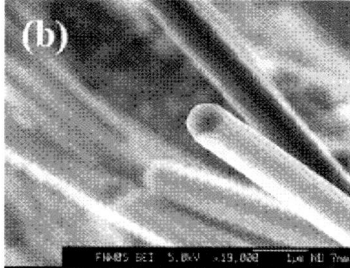

Figure 15.2 (a) SEM image showing typical C_{60} FNWs obtained by using the LLIP method. (b) Detailed view showing the typical cross section of the C_{60} FNWs. (Reprinted with permission from Ogawa et al.,[10] © 2006, American Institute of Physics.)

Typical C_{60} FNWs obtained by using the LLIP method were about less than 0.3 μm in diameter and greater than 100 μm in length (Fig. 15.2a). Also, these FNWs are generally straight, or only slightly curved, with an almost constant diameter along their growth axis. The smallest FNWs obtained using this method were found to be about 20 nm in diameter, as confirmed by AFM inspection. Both SEM (scanning electron microscope) and AFM inspections confirm that the FNWs exhibit a hexagonal cross section.

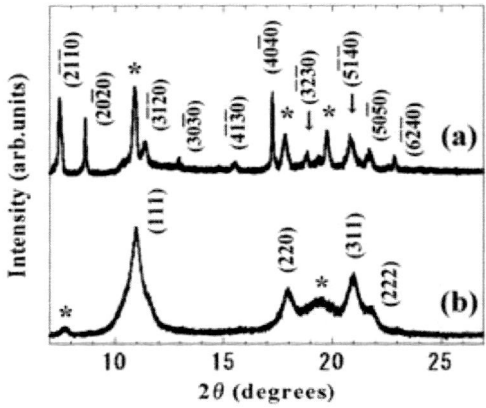

Figure 15.3 X-ray powder diffraction spectrum of (a) pristine C_{60} FNWs and (b) C_{60} FNWs after annealing at 440 K under about 2×10^{-6} Torr. Also identified in (b) are peaks arising from diffraction via specific crystal planes. These peaks are consistent with an fcc crystal structure with a lattice constant $a = 1.39$ nm. The lines marked (*) are of undetermined origin. (Reprinted with permission from Ogawa et al.,[10] © 2006, American Institute of Physics.)

X-ray diffraction spectra obtained for the pristine and annealed FNWs are shown in Fig. 15.3. Based on our determination of Miller indices for the main four peaks in their spectrum, it appears that the annealed FNWs have an fcc crystal structure similar to that of C_{60} single crystals. The lattice constant inferred for the annealed FNWs ($a = 1.39$ nm) is slightly shorter than reported for C_{60} crystals ($a = 1.41$ nm).[23] This decrease of lattice constant has also been observed in high-resolution transmission electron microscopy studies of C_{60} FNWs.[4] In the spectrum for the pristine NWs, on the other hand, there is clear evidence for the presence of some inclusions or different crystal structures in the FNW crystal structure.[24] The solvate formed from m-xylene and C_{60} has previously been reported to exhibit a hexagonal crystal structure ($P6_3$, $a = 2.37$ nm, $c = 1.00$ nm).[25] The diffraction spectrum of our pristine FNWs is also consistent with this hexagonal form (Fig. 15.2b), with Miller indices in agreement with the results reported in Ref. 21. Since the boiling point of *m*-xylene is 412 K, under normal pressure, we believe that almost adsorbed *m*-xylene in the FNWs should have been desorbed on annealing the pristine FNWs and that this may account for the change in the structural parameters of the crystal that we observe. However, a similar structural change of FNWs from the solvated structure to the fcc structure by the evaporation of solvent molecules in air was reported recently.[26]

The basic device operation of the C_{60} FNW-FETs (Fig. 15.1) is shown in Fig. 15.4 at a number of different back-gate voltages (V_g). While we do not observe any evidence for FET behavior in devices fabricated from pristine FNWs, when performed without any annealing, the transistor curves shown in Fig. 15.4 were obtained in a device realized using annealed C_{60} FNWs. Also, we conclude that the C_{60} FNW-FETs behave as n-channel enhancement-type FETs and a standard analysis allows us to estimate a resulting carrier mobility of 2×10^{-2} cm^2/(V s) by following the standard relation in a linear region applied to C_{60} TF-FET and other organic FET,[1] $I_{sd} = (\mu W C_0/L)(V_g - V_t)V_{sd}$, where W, L, V_{sd}, V_t, and C_0 are channel width, length, source–drain voltage, threshold voltage, and capacitance of the device, respectively. Here, $W = 15$ μm, $L = 4$ μm, and $V_t = \sim 0$ V for the device. The FNWs bridging the FET electrodes do not contact the SiO$_2$ layer directly, but, due to the thickness of these electrodes, are suspended at a height of about 10 nm above the substrate (this has been confirmed by AFM inspection). In our estimation for the carrier mobility, we have therefore assumed a compound electrostatic capacitance, with contributions due to the SiO$_2$ layer and the air gap. It is clear from the results of Fig. 15.4 that the C_{60} FNW-FET does not exhibit saturated behavior for any of the gate voltages shown. It should be pointed out here that our device may have a high contact resistance because of our simple approach of using FNWs to bridge the source and drain electrodes. In such a case, we expect that saturation was not observed since only a small fraction of the applied V_{sd} was actually dropped across the C_{60} FNWs.

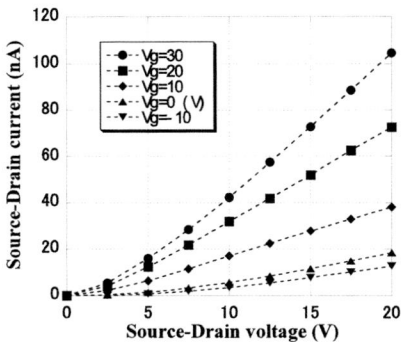

Figure 15.4 Transistor action of a vacuum-annealed C_{60} FNW-FET at room temperature. The FET exhibits n-channel enhancement-type behavior. (Reprinted with permission from Ogawa et al.,[10] © 2006, American Institute of Physics.)

Next, we have tried to fabricate another FET device using a single wire of C_{60} FNW.[11] From both SEM and AFM observations, the FNWs exhibit a hexagonal cross section that can be confirmed clearly in Fig. 15.5a. In the figure, the line profile of C_{60} FNW shows a flat at around the top due to the hexagonal cross section. To fabricate the C_{60} FNW-FET, we used a heavily doped (5×10^{-3} Ω cm) Si wafer as a substrate and as a back gate. This substrate had an insulating top layer of a thermally grown SiO_2 (500 nm), as shown in Fig. 15.5b. Source and drain electrodes of Au/Ti (20/10 nm) were fabricated on the surface of SiO_2 by photolithography and lift-off process. The device had a channel length of 5 μm, and a piece of the C_{60} FNW was used to bridge the two electrodes by means of a three-dimensional manipulator with a glass microcapillary under a high-resolution optical microscope. In order to remove intercalated oxygen and solvents in the C_{60} FNW, the device was annealed for about 24 h at ~440 K and ~10^{-6} Torr. After annealing, the FET characteristics have been measured without exposing to air and a gas exposure effect has been checked in the resistance of C_{60} FNW by introducing N_2 and O_2 gas into the vacuum chamber for measurements.

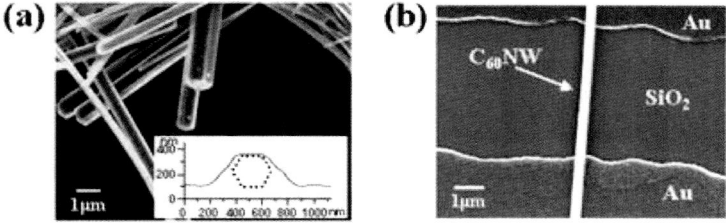

Figure 15.5 (a) SEM image showing typical C_{60} FNWs obtained by using the LLIP method. The inset shows the typical line profile of the C_{60} FNW obtained by AFM. The dotted line is guide for eyes. (b) SEM image showing the C_{60} FNW bridging between the two electrodes in the C_{60} FNW-FET device.[11]

Before annealing the C_{60} FNW-FET, no evidence of FET characteristic was observed. However, after 24 h of annealing at ~440 K and ~10^{-6} Torr, a clear FET operation was appeared. In several reports related to C_{60} FNWs, the FNWs have been shown to exhibit a hexagonal crystal structure due to intercalation of m-xylene molecules among C_{60} molecules.[24, 25] Recently, it has been reported that the crystal structure of C_{60} FNW changes to fcc structure by the evaporation of solvent (m-xylene) in air.[25] When annealing our device, solvent molecules (m-xylene) in the C_{60} FNW were evaporated as well. Therefore, after annealing, this device exhibited FET performance because of the evaporation of most solvent molecules and the consequent crystal-structure change. The C_{60} FNW still retains their shape without sublimation even at 440 K, since the usual sublimation temperature of C_{60} is higher than about 573 K under ~10^{-6} Torr. Figure 15.5 shows the SEM image of our C_{60} FNW bridging between the two electrodes. I_{sd} vs V_{sd} characteristics of the C_{60} FNW-FET at several V_g are shown in Fig. 15.6a, which shows n-type FET properties. The result of the gate-voltage characteristics ($I_{sd} - V_g$) is shown in Fig. 15.6b. The threshold voltage is about 10 V, and the field-effect mobility can be estimated to be about 2.5×10^{-2} cm^2/(V s).

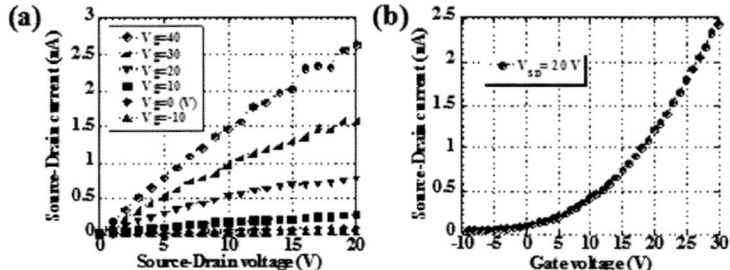

Figure 15.6 (a) I_{sd}–V_{sd} characteristics of the C_{60} NW-FET device at several gate voltages. (b) I_{sd}–V_g characteristics of the C_{60} NW-FET device at V_{sd} = 20 V.[11]

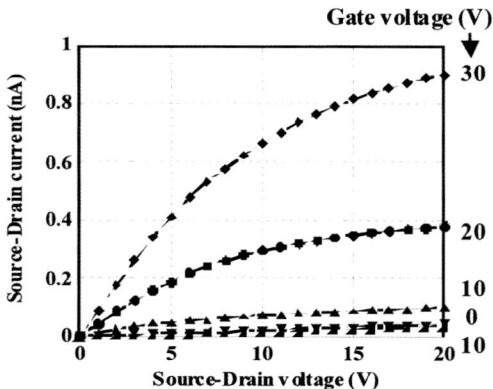

Figure 15.7 I_{sd}–V_{sd} characteristics at several gate voltages of the C_{60} FNWs-FET with the Mg electrode.[29,30]

In Fig. 15.6a, the I_{sd} does not exhibit saturation even up to V_g = 40 V. However, many papers have reported observation of a saturation region in C_{60} TF-FET.[3,13,19] One of the possible reasons is that the C_{60} FNW was put only on the electrodes, and therefore the contact resistance between the C_{60} FNW and the electrodes became inevitably high. This may play an important role in preventing saturation of our device. In order to improve the field-effect mobility of the C_{60} FNW-FETs, we must firstly reduce such a contact resistance. Probably, the use of low work-function metal electrodes would be effective to realize a good ohmic contact.[27] Moreover, we have observed a change of the resistance in annealed C_{60} FNW under N_2 and O_2 gas. We have introduced O_2 gas into the vacuum chamber just after introducing N_2 gas. Although the resistance is unchanged in the case of introducing N_2 gas, the resistance suddenly increases by 2 orders on the addition of O_2 gas. A similar behavior under gas purging has been observed in C_{60} TF-FET,[28] which also indicates that the electrical properties of C_{60} FNW are as influenced by oxygen absorption as C_{60} thin films.

Among the possible improvements of C_{60} FNW-FET characteristics, reduction of their contact resistance is needed for real applications. In usual gold contacts, the work function is 4.9 eV and the Fermi level is almost located at the middle of the HOMO–LUMO gap. However, for Mg contacts, the work function is 3.6 eV and it is just near the LUMO band. Indeed, in the case of C_{60} TF-FET using Mg metal contact as a source–drain lead contact, a clear improvement of the device performance can be obtained.[29] Therefore, the same improvement can be considered using low work-function metals for lowering the Schottky barrier in the lead contacts in C_{60} FNW-FET. In another application of top contact of the Mg electrodes for bundled C_{60} FNW-FET,[29] a large increase has been found in the source–drain current characteristics, which begin to show a saturated behavior for several gate voltages, as shown in Fig. 15.7. The use of the Mg electrodes thus suggests that a lower contact resistance is then obtained with the Au electrodes, as shown in Fig. 15.4. This is ascribed to the low work function of these electrodes, and the fact that they utilize a top-contact configuration. However, in the former two cases (Figs. 15.4 and 15.6), the C_{60} FNW was placed on top of the metal contacts without any covering. When we apply the top contact as a source–drain lead, the C_{60} FNW should be tightly fixed in mechanically and electrically. Therefore, by both using low work-function electrodes and introducing a top-contact configuration, a C_{60} FNW-FET device with Mg electrodes shows good performance in comparison with Au electrodes. Also, incidentally, the carrier mobility of the Mg-electrode device is slightly higher than that of the Au electrodes.

15.4 SOLVATED C_{60} FULLERENE NANOWHISKER FET

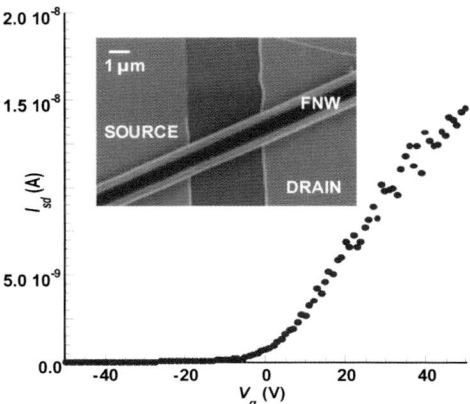

Figure 15.8 I_{sd}–V_g characteristic of a (*m*-xylene) C_{60} FNW-FET measured under the solvated condition. The inset shows a SEM image of the channel region of the FET.[31]

Here, we tried to operate the FET under ambient conditions. In Fig. 15.8, we show the V_g characteristic of a C_{60} FNW-FET, which was obtained by using the conducting substrate of the device as a back gate (this accounts for the large values of this voltage). The LLIP process used to synthesize the FNW in this case involved the use of *m*-xylene solvent, and the measurements of Fig. 15.8 were obtained under the solvated condition due to the presence of a N_2 atmosphere. From the V_g dependence of the I_{sd}, it can be inferred that the V_t and field-effect mobility are ~0 V and 4 × 10^{-3} cm²/(V s), respectively (we determine the field mobility as the same as already mentioned). The device shows a normal-on characteristic similar to our former reports, with a mobility only a little lower than that which we have previously found[10] for dried FNW-FETs [2.5×10^{-2} cm²/(V s)]. This represents an important result, since it demonstrates the possibility of obtaining FET action from FNWs under other than vacuum conditions.

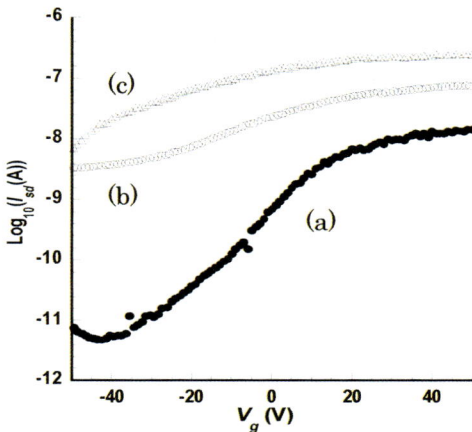

Figure 15.9 $I_{sd}-V_g$ characteristic (with current plotted on a logarithmic scale) of the FNW-FET, measured at room temperature under three different conditions: N_2 atmosphere under the solvated condition (curve a, filled circles), dried in vacuum (curve b, open circles), and dried in vacuum and annealed (curve c, open triangles).[31]

Subsequent to the measurements of Fig. 15.8, the probe-station chamber was next evacuated to dry out the FNWs. After measuring the resulting FET characteristics, the device was then annealed at 440 K for 24 h and cooled back down to RT. The resulting gate-voltage characteristics for these conditions are shown in Fig. 15.9. Note that the current in this figure is plotted on a logarithmic scale, for the solvated (curve a), dried (curve b), and dried and annealed (curve c), conditions. I_{sd} ($V_g = 0$) increases on evaporating the solvent, while the on/off ratio decreases from 4 to 2 orders of magnitude. On the other hand, we find that the mobility is improved to 10^{-2} cm^2/(V s) as a result of the annealing, an improvement of roughly a factor of 2 over the solvated condition.

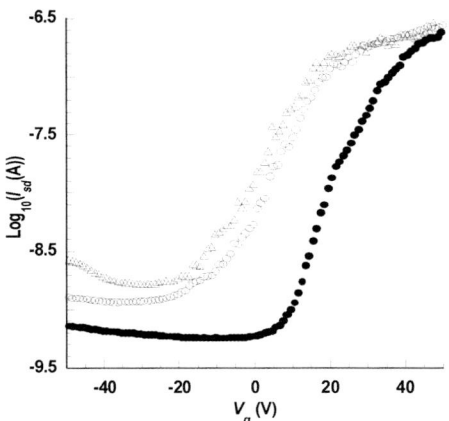

Figure 15.10 I_{sd}–V_g characteristic (with current plotted on a logarithmic scale) of a (chlorobenzene) C_{60} FNW-FET, measured at room temperature under three different conditions: N_2 atmosphere (solvated condition, filled circles), dried in vacuum (open circles), and dried in vacuum and annealed (open triangles).[31]

Also the same those procedures have been performed as in another FET whose channel was formed from FNWs realized using chlorobenzene as the solvent. The corresponding variations of I_{sd} (V_g) measured in this case are shown in Fig. 15.10. From these data, we are again able to infer that the solvated FNW yields transistor characteristics that compare favorably with those obtained under vacuum conditions. The on/off ratio is a factor of 10 larger in the solvated condition, but the inferred mobility is actually slightly higher.

In the solvated condition, it appears that the mobility of the chlorobenzene-derived FNW is significantly enhanced over that of the *m*-xylene-derived one. Although the reasons for this are not completely understood, an important factor may be the different polar nature of the two solvents we have used. Specifically, Cl should act as a weak acceptor in chlorobenzene, while CH_3 acts as a donor in *m*-xylene with benzene. These results therefore suggest the importance of utilizing an appropriate solvent as a means to enhance the electrical characteristics of the solvated FNWs. The other noteworthy feature concerns the characteristics achieved after drying and annealing the FNWs. Since in this case the solvent is removed from the FNW, one might expect to obtain the same electrical characteristics for FNWs produced with either solvent. It indicates that the mobility and on/off ratios are the same in both cases. However, the threshold voltages are different; there exists an unintentional doping induced by the different solvents.

Turning to the physical implications of our results, the observation of significant changes in the threshold voltage of both devices indicates that carriers are introduced into the FNWs, independent of the gate voltage, by evaporating the solvent. This may possibly be related to the introduction of a high density of dislocation and disorder,[9] which could result in the creation of additional carriers in the FNWs. The low values of the mobility that we infer for both devices in the solvated condition appear reasonable, particularly in the case of m-xylene, since solvent molecules intercalated into the C_{60} crystal should extend the lattice constant, thereby decreasing the carrier hopping probability.[17] It is clear, on the other hand, that the mobility is not changed so systematically by removing the solvent from the FNWs. This may indicate some competition between the effect of eliminating the intercalated solvent molecules and the ensuing increase in crystalline disorder when the crystalline structure of the FNWs undergoes a transition to the fcc form.

15.5 SUMMARY

It was discussed here that C_{60} synthesized by using the LLIP method can be utilized as the channel of three-terminal FET structures. The fcc crystal structure of annealed FNWs was found by X-ray diffraction to be similar to that of C_{60} crystals, although with a slightly reduced lattice constant. This indicates that the transport properties of C_{60} are directly connected to those of C_{60} FNWs and give important information on fabrication of FNW-FETs. The FNW-FETs exhibited n-channel enhancement-type transport behavior, and their carrier mobility was estimated to be 2×10^{-2} cm^2/(V s) under vacuum conditions at room temperature.[10] However, the operation of C_{60}

FNW-FETs under a N_2 environment has been demonstrated, allowing reasonable transistor action to be achieved without the need to operate devices under vacuum. The crystalline structure of the FNWs changes from an oriented hexagonal form to a disordered fcc one on evaporating the solvent. Accompanying this, resulting dislocations in the crystalline structure are believed to introduce carriers into the FNWs, thereby modifying the threshold gate voltage. The finding that solvated FNWs may be used to implement FETs could have important implications by allowing for technological applications of these materials to be developed. Future work should therefore focus on exploring suitable solvents for FNW fabrication to yield optimal electrical characteristics.

Acknowledgments

The authors acknowledge the contributions to the work presented here by all their past and present collaborations, especially Dr. K. Miyazawa, Prof. M. Tachibana, Dr. T. Sasaki, Dr. K. Horiuchi, Mr. T. Kato, Mr. K. Ogawa, Mr. H. Tsuji, Mr. Y. Chiba, Mr. S.-R. Chen, and Prof. J. Onoe. This work was supported in part by a Grant-in-Aid for Scientific Research from the Japan Society for the Promotion of Science (Nos. 19054016, 16656007, and 16206001). Also, the work was in part supported by Global COE program at Chiba University (G-03, MEXT) and JSPS Core To Core Program.

References

1. C. D. Dimitrakopoulos, and P. R. L. Malenfant, *Adv. Mater. (Weinheim, Ger.)*, **14**, 99 (2002).
2. T. D. Anthopoulos, B. Singh, N. Marjanovic, N. S. Sariciftci, A. M. Ramil, H. Sitter, M. Cölle, and D. M. de Leeuw, *Appl. Phys. Lett.*, **89**, 213504 (2006).
3. K. Horiuchi, K. Nakada, S. Uchino S. Hashii, A. Hashimoto, T. Sasaki, N. Aoki, Y. Ochiai, and M. Shimizu, *Appl. Phys. Lett.*, **81**, 1911 (2002).
4. K. Miyazawa, A. Obayashi, and M. Kuwabara, *J. Am. Ceram. Soc.*, **84**, 3037 (2001).

5. K. Miyazawa, Y. Kuwasaki, A. Obayashi, and M. Kuwabara, *J. Mater. Res.*, **17**, 83 (2002).
6. F. Michaud, M. Barrio, S. Toscani, D. O. Lopez, J. L. Tamarit, V. Agafonov, H. Szwarc, and R. Ceolin, *Phys. Rev. B*, **57**, 10351 (1998).
7. S. Ogawa, H. Furusawa, T. Watanabe, and H. Yamamoto, *J. Phys. Chem. Solids*, **61**, 1047 (2000).
8. P. R. L. Malenfant, C. D. Dimitrakopoulos, J. D. Gelorme, L. L. Kosbar, T. O. Graham, A. Curioni, and W. Andreoni, *Appl. Phys. Lett.*, **80**, 2517 (2002).
9. K. Miyazawa, J. Minato, T. Yoshii, M. Fujino, and T. Suga, *J. Mater. Res.*, **20**, 688 (2005).
10. K. Ogawa, T. Kato, A. Ikegami, H. Tsujii, N. Aoki, Y. Ochiai, and J. P. Bird, *Appl. Phys. Lett.*, **88**, 112109 (2006).
11. K. Ogawa, T. Kato, A. Ikegami, H. Tsuji, N. Aoki, J. P. Bird, and Y. Ochiai, *J. Phys.: Conf. Ser.*, **39**, 1 (2006).
12. J. Paloheimo, H. Isotalo, J. Kastner, and H. Kuzmany, *Synth. Met.*, **55–57**, 3185 (1993).
13. R. C. Haddon, A. S. Perel, R. C. Morris, T. T. M. Palstra, A. F. Hebard, and R. M. Fleming, *Appl. Phys. Lett.*, **67**, 121 (1995).
14. K. Kaneto, K. Yamanaka, K. Rikitake, T. Akiyama, and W. Takashima, *Jpn. J. Appl. Phys.*, **35**, 1802 (1996).
15. B. Pevzner, A. F. Hebard, and M. S. Dresselhaus, *Phys. Rev. B*, **55**, 16439 (1997).
16. K. Horiuchi, S. Uchino, K. Nakada, N. Aoki, M. Shimizu, and Y. Ochiai, *Physica B*, **329–333**, 1538 (2003); K. Horiuchi, S. Uchino, K. Nakada, N. Aoki, M. Shimizu, and Y. Ochiai, Proceedings of the 26th International Conference on the Physics of Semiconductors, *Inst. Phys. Conf. Ser.*, **171** (2003).
17. K. Horiuchi, S. Uchino, S. Hashii, A. Hashimoto, T. Kato, T. Sasaki, N. Aoki, and Y. Ochiai, *Appl. Phys. Lett.*, **85**, 1987 (2004).
18. Y.-Y. Lin, D. J. Gundlach, S. F. Nelson, and T. N. Jackson, *IEEE Electron Device Lett.*, **18**, 606 (1997).
19. S. Kobayashi, T. Takenobu, S. Mori, A. Fujiwara, and Y. Iwasa, *Appl. Phys. Lett.*, **82**, 4581 (2003).

20. Y. S. Yang, S. H. Kim, J. Lee, H. Y. Chu, L. M. Do, H. Lee, J. Oh, and T. Zyung, *Appl. Phys. Lett.*, **80**, 1595 (2002).
21. T. Kanbara, K. Shibata, S. Fujiki, Y. Kubozono, S. Kashino, T. Urisu, M. Sakai, A. Fujiwara, R. Kumashiro, and K. Tanigaki, *Chem. Phys. Lett.*, **379**, 223 (2003).
22. K. Horiuchi, T. Kato, S. Hashii, A. Hashimoto, T. Sasaki, N. Aoki, and Y. Ochiai, *Appl. Phys. Lett.*, **86**, 153108 (2005).
23. P. A. Heiney, J. E. Fisher, A. R. Mc Ghie, W. J. Romanov, A. M. Denestein, J. P. McGauley, A. B. Smith, and D. E. Cox, *Phys. Lett.*, **66**, 2911 (1991).
24. Z. Otto and E. C. David, *J. Phys. Chem. Solids*, **53**, 1373 (1992).
25. M. Ramm, P. Luger, D. Zobel, W. Duczek, and J. C. A. Boeyens, *Cryst. Res. Technol.*, **31**, 43 (1996).
26. J. Minato and K. Miyazawa, *Carbon*, **43**, 2837 (2005).
27. M. Chikamatsu, S. Nagamatsu, T. Taima, Y. Yoshida, N. Sakai, H. Yokokawa, K. Saito, and K. Yase, *Appl. Phys. Lett.*, **85**, 2396 (2004).
28. A. Tapponnier, I. Biaggio, and P. Gunter, *Appl. Phys. Lett.*, **86**, 112114 (2005).
29. K. Ogawa, A. Ikegami, H. Tsuji, T. Kato, N. Aoki, J. P. Bird, and Y. Ochiai, *Proceedings of the 28th International Conference on the Physics of Semiconductors, Vienna, July 2006*.
30. K. Ogawa, N. Aoki, K. Miyazawa, S. Nakamura, T. Mashino, J. P. Bird, and Y. Ochiai, *Jpn. J. Appl. Phys.*, **47**, 501 (2008).
31. Y. Ochiai, K. Ogawa, N. Aoki, and J. P. Bird, *J. Phys.: Conf. Ser.*, **159**, 012004 (2009).

Index

1-butanol   78–79
2-butanol   94
2-propanol   64, 66–70, 72, 91, 93–95, 139, 187–188, 194

absorption edge   149–151
ACWs *see* amorphous carbon whiskers
AFM *see* atomic force microscopy
AFM images   172–173, 175
AFM imaging   169, 172
air-drying   32, 34
alcohols   4, 8, 10, 26–27, 64–66
   vinyl   137, 139, 141–142
aldehydes   56
aligned C_{60} microtubes   10, 68–72
alumina   210
amorphous carbon whiskers (ACWs)   109–110, 114
amplitude modulation   169
annealing   156–157, 159–160, 212, 216–219, 223, 225
as-grown C_{60} nanowhiskers   199, 201
as-prepared C_{60} nanowhiskers   204
atomic force microscopy   14, 104, 119, 163–165, 167, 171–172, 180, 214, 219
Auger maps   175–176
azomethine ylides   55–56

bipyramid C_{60} 97
buckling test   108, 119, 122–123
bulk C_{60} crystals   91, 118, 123

bulk crystals
   fcc C_{60} 34–35
   hexagonal C_{60} 34
bulk fullerene crystals   90

C_{60} 25–26, 28, 30, 32, 34, 36, 38, 40, 42–44, 46, 89–90, 139–143, 174–175, 189, 209–222, 224–225
   condensation of   31, 147–148
   double bonds of   55–56
   mechanical characterization of   118–119
   microbelt   96–97
   nano-bipyramid   97
   optical properties of   147–148
   reaction of   55–56
   saturated pyridine solution of   139
   shape formation process of   95–96
   sublimation of   198–199
   vibronic coupling of   147–149
C_{60} nanowhiskers
   aligned   140
   fabricated   90
   incorporated   83
C_{60} radical anions   30
C_{60}NWs   45, 47, 49, 158, 191
   grown   29–31
   growth rate of   7, 30, 51
   length growth rate of   45, 47
   length of   43, 45, 49, 51
   modified   38
   pressed   187, 189–191
cantilever bend test   131
carbon atoms   56, 118, 191–192, 194

carbon nanocapsules 103, 109, 111
carbon nanomaterials 60, 103–104, 114
  low-dimensional 19, 54
carbon nanotubes 40, 54, 77, 113, 118, 138–139, 206, 210
Ce-ion-incorporated fullerene 81
Ce-ion-incorporated nanowhiskers 82–83
Ce ions 78, 82–83
cellulose fibers 139
chemical bonds 118
chemical calculations 154
chemical inhomogeneity 168–170
chemical mapping 164
chlorobenzene 224
conductance 103–104, 111–114, 210, 212–213
conductivity 77, 175, 177
crystal field 35
crystal structure 26, 32, 83, 95, 97, 186, 211, 216, 219
  solvated 142
crystalline structure 17, 225–226
crystallinity 152, 156, 158–160
crystals 3, 5, 26–27, 31, 35, 57, 63–64, 67, 72–73, 91, 94–95, 152–154, 178, 211, 216–217, 225
  molecular 26
  seed 95
  single 3, 7, 19, 32, 104, 122, 151–153, 211, 216
  solution-grown 39
  solvent-free 26
Curie–Weiss law 189–190
curved nanowhiskers 28
cyclopropanation 55

derivative molecules 3, 13, 57
derivatization 54
diethyl bromomalonate 55

direct methanol fuel cells (DMFCs) 81
DMFCs *see* direct methanol fuel cells
dried C_{60}NWs 14, 185–188, 193–194

elastic deformation 25, 37, 39
electrical conductivity 77, 138, 174, 204
electron-beam (E-B) lithography 210
electron beams 7, 17, 19, 186
evaporation 142, 219

Fe-ion-incorporated fullerene 78
Fe-ion-incorporated nanowhiskers 83
Fe-ion incorporation 78–79
Fe ions 78–79
FET characteristics 215, 218–219, 223
FETs *see* field effect transistors
FFM *see* friction force microscopy
field-effect mobility 219–220, 222
field effect transistors (FETs) 16–17, 174, 210–215, 218, 220, 222, 224, 226
fine crystals 89–100
  inorganic 91
  monodispersed 95
  solvated 95
FNWs
  air-exposed 176
  annealed 213, 216, 225
  annealed C_{60} 221
  chlorobenzene-derived 224
  fibrous C_{60} 211
  hexagonal 13
  individual C_{60} 174
  insulator-shelled 176
  precipitated 211

prepared 140
pristine 110, 216–217
single-crystal C_{60} 214
solvated 224–226
solvated C_{60} 211
surface layer 180
force–deflection characteristics 126, 132
force–deflection–power characteristics 126, 132
fracture 109, 123–124, 126, 129, 131
fractured C_{60} nanowhisker 129
friction force microscopy (FFM) 168
fullerene-based materials 63
fullerene-based nanostructures 77
fullerene-based solar cells 86
fullerene C_{60} crystals 138
fullerene C_{60} nanowhiskers 138–139
fullerene chemistry 118
fullerene crystals 4
fullerene derivatives 19, 53–58, 60–62, 77
  incorporation of 53, 60
  pyrrolidine-type 61
  ratio of 58, 60
  typical 53
  water-soluble 55
fullerene derivatization 56
fullerene nanobelt 4
fullerene nanofibers 4, 19, 197–198, 200, 202, 204, 206
  dried 198
  heat-treated 19
fullerene nanomaterials 4, 19, 77, 163–164
  low-dimensional 1
fullerene nanoparticles 170–171, 177
fullerene nanoribbon 4
fullerene nanorod 4
fullerene nanosheets 4, 76, 84–86
  flexible shape 86
  polygonal 85–86
  rhombus shape 86
  tunable hexagonal 85
fullerene nanostructures 75–76, 78–79, 84, 86
  metalion-incorporated 78
  precipitation of 76, 78
  preparation of metal-ion-incorporated 78, 84, 86
fullerene nanotubes 1, 3, 19, 77, 138, 164, 198
  crystalline 109
fullerene nanowhiskers 1–4, 6, 10–12, 18–20, 43–44, 52–54, 75–80, 82, 103–104, 116–118, 137–138, 146–148, 162–164, 185–186, 197–198, 209–212
  incorporated 83
  isolated crystalline 103
  metal-free nanoporous 80
  nanoporous 80
  nontubular 198
  optimal metal-ion-incorporated 84
  porous Ce-ion-incorporated 81
  single crystalline 138
  tubular 198
  tubular nanoporous 81
fullerene nanowires 4, 138
fullerene sheets 85
fullerene solutions 10
fullerenes 3–4, 13, 43, 54, 56–59, 63–65, 77–79, 84, 89–91, 95, 138, 147, 174, 198, 210, 212
  endohedral 3
  incorporated 80
  mechanical bend testing of 117, 135
  modified 63
  multiwall 205
  multiwalled 110
  spherical 90
fullerenols 54

g-parity 151, 154
gate voltages 217, 220–221, 225
gold nanoparticles 90, 96, 98
grain boundaries 118, 134, 212
graphitic ribbons 197, 203–204, 206
graphitization 159–160
   photo-induced 160

heat-treated C_{60} nanowhiskers 199, 203–204
Hertzberg–Teller mechanism 149, 151, 153
hexagonal nanowhisker 28
high-resolution transmission electron microscopes 199, 206
hysteresis 190

injection mode 70–71
injection rate 10, 66–69
intensity 158–160, 169, 192–193
   emission 110
   integrated 154–155
   relative 31, 155, 158–159
interlayer spacing 197, 204, 206
   average 204
intermolecular interaction 33, 35, 91, 95
irradiation 5, 17–18, 186
irradiation time 158–159
isopropyl alcohol 4, 28, 43–44, 57, 78, 119, 187, 214

Jahn–Teller mechanism 151, 153

KBr crystals 18
   needlelike 18–19

laser irradiation 158–159
laser irradiation time 158–160
lateral force microscopy (LFM) 168
lattice constants 5, 7, 11–12, 14, 185, 187–189, 216, 225
LFM *see* lateral force microscopy
liquid–liquid interface 4, 7, 10, 27–28, 44, 214
liquid–liquid interfacial precipitation (LLIP) 1, 3, 5, 25, 27, 44, 57, 63, 75–76, 91, 118–119, 138, 148, 185–186, 198, 211
liquid–liquid interfacial precipitation method 1, 3, 25, 44, 75, 91, 138, 148, 185, 198
LLIP *see* liquid-liquid interfacial precipitation
LLIP process 64–65, 222
loaded C_{60} nanowhisker 134
luminescence 103, 110

magnetic alignment 137, 139, 141–142
magnetic bars 45–46
magnetic fields 137, 141–142, 187
magnetic susceptibility 185, 187, 189
malonic acid derivatives 55, 60
metallic nanotips 111–112
methanol soaking 156
microchannel 10–11
microrods 70
Miller indices 216–217
molecular reorientation 194

nanoballs 92, 94
nanocrystals 91, 96, 147–148
nanofibers 19, 44, 197–199, 204, 206
nanomaterials 61, 104
nanoscale devices 25, 27, 39
nanoscaled phases 147–148, 153, 155, 157

nanotechnology 22, 148, 182, 212
nanowhiskers 1–2, 25–28, 42–43,
    53–54, 58–60, 62–64, 77,
    81–84, 108, 118–120,
    122–124, 128–130, 137–139,
    199, 201–202, 204–206
  brittle 129
  isolated 28
  long 123
  metalion-incorporated 83
  porous 82
Ni-ion-incorporated C_{60} 80–81
NMR *see* nuclear magnetic
    resonance
NMR measurements 185, 187–
    188, 191, 193–194
nuclear magnetic resonance (NMR)
    185–186
nucleation rate 73

optical absorption 31, 150–151
  broad 30
organic materials 40, 64, 165, 214
oscillation amplitude 168–169
oxygen 2, 176, 180, 210–211
oxygen intercalation 176–177, 180

pericyclic reactions 55
phase 168–169, 186–187
  graphitic 147–148
  low temperature 155
  perovskite 2
phase imaging 163, 169, 180
photo-induced structural
    transformations 147–148,
    160
photopolymerization 5, 29,
    158–160
piezomanipulation 105, 111, 114
platinum catalysts 206
polymerization 18, 83–84,
    123–124, 159, 178, 186
  photo-induced 158, 160

precipitator injection rate 66–67
pristine C_{60} 187–194
pristine C_{60} crystals 77, 155, 194,
    198
pristine C_{60} FNWs 175, 216
pristine C_{60} powder 77, 83,
    185–191
proton decoupling 192
pulverization 141
pyridine solution 7–9, 14
pyridine solution of C_{60} 12
pyrochlore phase 2
pyrrolidine derivatives 55–56

quantized conductances 112,
    114

radical anions 30–31
Raman bands 33–34
Raman spectra 31, 34, 75, 83–84,
    158, 199, 201
Raman spectrum 158–159, 199
Raman spectrum of as-grown C_{60}
    nanowhiskers 199
ratios
  molar 95
  on/off 223–225
  solvent 9, 47
  volume 9, 64
reprecipitation 91–93, 95
resistivity 15–16, 77, 110, 204
  electrical 15–16, 110, 198,
    204
resistivity change 15–16
resonance frequency 168–169,
    177

SAEDP *see* selected-area electron-
    diffraction pattern
SAM *see* scanning Auger microscopy
scanning Auger microscopy (SAM)
    164

scanning electron microscopy (SEM) 2, 28, 44, 48, 69, 90, 97–98, 122, 140, 156, 164, 198, 215
scanning force microscopy (SFM) 163, 165
selected-area electron-diffraction pattern (SAEDP) 5, 8, 14, 17–19, 58, 60, 199–200
SEM *see* scanning electron microscopy
semiconductors 76, 91, 182, 227–228
sensors 2, 119–120, 165
SFM *see* scanning force microscopy
SiO_2 177, 214–215, 218
SiO_2 layer 217
sodium hydride 55
solubility 57, 78, 84
solution growth 25–26, 39
solution temperature 4
source–drain voltage 217–218, 220
sp^2 character 192, 194
SPM (scanning probe microscopy) 163–165, 179
stick diagram 154
STM (scanning tunneling microscopy) 104, 164–165
sublimated C_{60} crystals 72
sublimation 34–35, 178, 219
supersaturation 9, 51, 64, 66–68
supersaturation of C_{60} 65–66
surface energy 73
surface topography 172

TEM *see* transmission electron microscopy
temperature dependence 47, 155, 185, 188–191
temperatures 2, 7, 9, 17, 43–45, 64, 75–76, 79, 84, 91, 93–94, 153, 155, 158, 178, 187
  aging 91–92

average device 127–128
elevated 164, 177–178
high 17, 84, 118, 188, 198, 201, 204, 206
low 155, 188, 192
tensile loading 119–121
tester actuation 130
time constants 159–160
time evolution 31–34, 158–159
tip–surface interactions 165, 167
toluene 4, 12, 19, 44, 46, 48, 51, 54–55, 57, 64, 66, 70, 78–79, 149–151, 187–188, 192–194
toluene solution 4, 10, 26–28, 30, 34–35, 43–44, 47, 49, 187
torsional deflection 166, 168
transition metal ions 77
transmission electron microscopy (TEM) 4, 14, 17, 58, 80, 90, 92, 104, 119, 123, 140, 164, 186, 197–198
tunneling current 165

ultrahigh vacuum 169, 177–178
ultrasonication 9–10, 19, 44

vacuum 17, 38, 77, 104, 106–107, 109–110, 112, 148, 153, 158, 178, 197, 204–206, 210, 223–224, 226
vacuum chamber 218, 221
vacuum conditions 211, 213, 222, 224–225
vacuum-drying 32
variable range hopping (VRH) 212–213
voltages 109, 220, 222
  acceleration 17
  threshold 217, 219, 225
volume mixing ratios 43, 47–49, 51

VRH *see* variable range hopping

water   7, 43, 49–50, 139
   distilled   49, 64, 68
water content   43–44, 49, 51–52
whisker diameter   133
wind pressure   35–38
work function   221

X-ray diffraction (XRD)   31, 83, 95,
   186, 214, 225
XRD *see* X-ray diffraction
XRD pattern   31–32, 84, 99

xylene   4, 12, 33, 57, 78–79, 91,
   93–95, 156, 214, 216–217,
   219, 222, 224–225
xylene molecules   156, 219
xylene solutions   11, 33, 93–94,
   214

Young's modulus   14, 37, 106–108,
   117–119, 122–123, 126,
   132–135

zero-field-cooled (ZFC)   189–190
ZFC (zero-field-cooled)   189–190
zirconate titanate   1–2